GLOBAL WARNING

GLOBAL

WARNING

The Last Chance for Change

PAUL BROWN | Foreword by **Gerd Leipold,** *Executive Director*
GREENPEACE INTERNATIONAL

Reader's Digest

The Reader's Digest Association, Inc.
Pleasantville, NY / Montreal / Singapore

A READER'S DIGEST BOOK

This edition published by The Reader's Digest Association, Inc.,
by arrangement with dakini books NP

Conceived and created by
www.dakinibooks.com

FOR DAKINI BOOKS
Publisher: Lucky Dissanayake
Sub-Editor: Paul Olive
Advisors: James Cameron, Jon Gibbons, Gerard Wynn, Phil Wolski, Hélène Connor
Researcher: Julia Zumstein
Initial Design: FORM
Reproduction: Richard Deal—Wyndeham Prepress

The carbon emitted by dakini books NP during the initial production
of this book has been offset.

FOR READER'S DIGEST
U.S. PROJECT STAFF
Project Editor: Kim Casey
Copy Editors: Barbara Booth, Mary Connell, Fiona Hunt
Indexer: Andrea Chesman
Associate Art Director: George McKeon
Project Production Coordinators: Wayne Morrison, Gretchen Schuler

CANADIAN PROJECT STAFF
Project Editor: Pamela Johnson
Assistant Editor: Jim Hynes
Contributing Editor: Camilla Cornell

Executive Editor, Trade Publishing: Dolores York
Director of Production: Michael Braunschweiger
Production Manager: Elizabeth Dinda
Associate Publisher: Rosanne McManus
President and Publisher, Trade Publishing: Harold Clarke

Library of Congress Cataloging in Publication Data

Brown, Paul, 1944-
  Global warning: the last chance for change / Paul Brown.
     p. cm.
  Includes index.
  ISBN 0-7621-0876-2 (978-0-7621-0876-3)
  1. Global warming. 2. Global environmental change. I. Title.
QC981.8.G56B763 2007
363.738'7--dc22
                    2007021808

SPECIAL THANKS

Many thanks to the following organizations and photographers,
who kindly donated the images featured in this book:

NASA/ESA/Greenpeace/WWF/Amjad Abdulla/George Fischer/
Brian Knutsen/Andy Harvey/Alastair Dobson/Rob Webster/
John N. Newby/Soh Koon Chng/Mario Farinato/His Excellency
Mr. Maumoon Abdul Gayoom, President of the Republic of Maldives/Coral Cay
Conservation Trust/Grant Gilchrist/Kenya Tourist Board/Scottish Power/Paul
West/Transport for London/European Commission/BP plc/Chevron/Oddgeir
Karlsson/Stirling Energy Systems/Urban Mover/General Motors/Kisho
Kurokawa/Kiss&Cathcart Architects/FxFowle Architects/Foster and Partners/Lifschutz
Davidson Sandilands/Honda/Shell/Seafish/FARMA

We are committed to both the quality of our products and the service we provide
to our customers. We value your comments, so please feel free to contact us.
The Reader's Digest Association, Inc.
Adult Trade Publishing
Reader's Digest Road
Pleasantville, NY 10570-7000

For more Reader's Digest products and information, visit our website:
www.rd.com (in the United States)
www.readersdigest.ca (in Canada)
www.rdasia.com (in Asia)

This book is printed on environmentally friendly paper certified
by FSC (Forest Stewardship Council).

MIXED SOURCES
Product group from well-managed forests and other
controlled sources. Certificate number SGS-COC-2659
FSC

Printed in China

1 3 5 7 9 10 8 6 4 2

"How could I look my grandchildren in the eye and say I knew about this and I did nothing?"

—Sir David Attenborough
Naturalist and Veteran Broadcaster

Contents

Foreword

The publication of this book comes as the future of the planet hangs in the balance. The recent conclusions from the IPCC (Intergovernmental Panel on Climate Change) have put the debate beyond any doubt: Climate change is real, and as the vivid images (some provided by Greenpeace) in the following pages show, its impact is already being felt around the world.

The main cause of global warming is simple: excessive emission of greenhouse gases due to human activities. But the causes of that carbon pollution are complex and varied. Carbon dioxide in the atmosphere now stands at its highest level in 650,000 years—already far higher than the level that wiped out entire species and changed whole continents, long before mankind—and these levels are rising more steeply than at any point in history.

As you look through these dramatic depictions of the causes and effects of climate change, you will realize just how much we all have to change. In fact, every reader of this book can have a powerful impact. As consumers, we can shift the status quo every time we make a decision to buy a sustainable, locally produced product; every time we avoid using our car or refrain from unnecessary flying; every time we use energy smartly by switching to efficient lightbulbs and save energy elsewhere in the home; every time we persuade someone else to do the same.

But that is just the start. Large-scale changes need to be instigated by governments. We must hold governments responsible for their inaction on climate change and force them to live up to their responsibilities—to invest in secure, cost-effective, clean renewable energy and to help disassemble the mechanisms that have allowed dirty, carbon-polluting power to thrive for so long.

To avoid the worst effects of global warming, we must keep the mean global temperature increase below 3.6°F (2°C). To achieve this, we must dramatically reduce and ultimately eliminate our use of fossil fuels as our primary source of energy, and we must generate a massive uptake of renewable energy coupled with energy efficiency.

What we need is a complete transformation in the way we produce, consume, and distribute energy—a radical departure in the way we think about energy. We need a revolution—one that achieves equilibrium between our energy consumption and the natural systems that sustain us—a revolution that ensures equitable access to the manifest benefits of electricity. Nothing short of such an Energy Revolution will enable us to limit global warming to less than 3.6°F (2°C).

We must learn once again to respect the planet's natural limits. Each year, we pump about 11 billion tons of man-made carbon dioxide into the atmosphere. The Earth may contain enough fossil fuels to supply our needs for perhaps a thousand years or more, but in the earth is where it must stay. There is only so much carbon that natural systems can absorb, and we must maintain and protect the health of our natural systems, such as the great rain forests of the Amazon, Africa, and Southeast Asia, as well as the planet's marine ecosystems.

We have only a decade—two at most—to reverse the trend, to radically reduce greenhouse gas emissions, and to avoid the worst impacts of climate change.

If you have ever been in awe of the power of Earth's climate and weather systems, then be inspired to do something for climate change… and do it today.

—Gerd Leipold, *Executive Director*
GREENPEACE INTERNATIONAL

Our Changing World

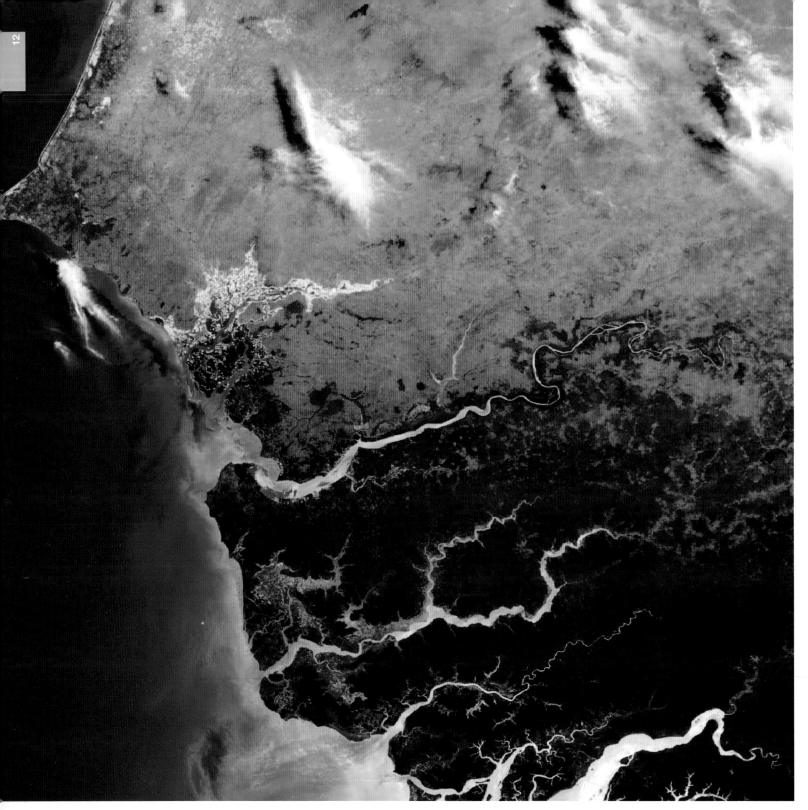

Previous spread: The permanent bright lights of Earth as seen from space. This remarkable image constructed from satellite photographs shows the most urbanized areas of the planet but not necessarily the most populated. Compare China and India with western Europe. The outlines of a world map have been superimposed for guidance; it is startling how many centers of population are situated on the coast and are vulnerable to sea-level rise. It is clear that the United States, Europe, and Japan are using large amounts of electricity for lighting cities at night. Note: The bright squiggle in eastern North Africa shows the concentration of population along the Nile, and across Russia the light marks the route of the Trans-Siberia Railway from Moscow to Vladivostok. Dark areas are the great deserts in Africa, Arabia, Australia, Mongolia, China, the United States, and the mountainous areas of the Himalayas. Although Iceland can be seen, Greenland and Antarctica remain dark.

Above: This picture covers the transition zone between the desert and savannah in the north of Africa and the tropical vegetation farther south. Clearly seen is the discharge into the sea of sediment via the river system along the west coast. This is the result of soil erosion, partly caused by overgrazing and the loss of vegetation. A series of satellite images over time can keep track of changes and pinpoint where intensification of land use is causing problems and where action is needed on the ground to halt the erosion.

Above: The Sahara Desert will spread north and leap the Mediterranean, according to scientists studying climate change. This may sound fantastic, but as this picture taken on November 14, 2004, shows, vast clouds of dust can be carried out over the sea from North Africa. This image shows Libya (center) and Tunisia (top left). Most of the dust fell in Europe over several days as a large weather system whipped up gale-force winds that affected Algeria, Italy, and as far north as Albania. Arid conditions in southern Spain, Portugal, Italy, and Greece were already causing concern before the prolonged drought and heat wave in 2005. All four countries are vulnerable to climate change, which will bring extra heat and prolonged droughts in this region. Because of the deteriorating conditions, these southern European states have all joined the U.N.'s Desertification Convention.

2002

2003

2004

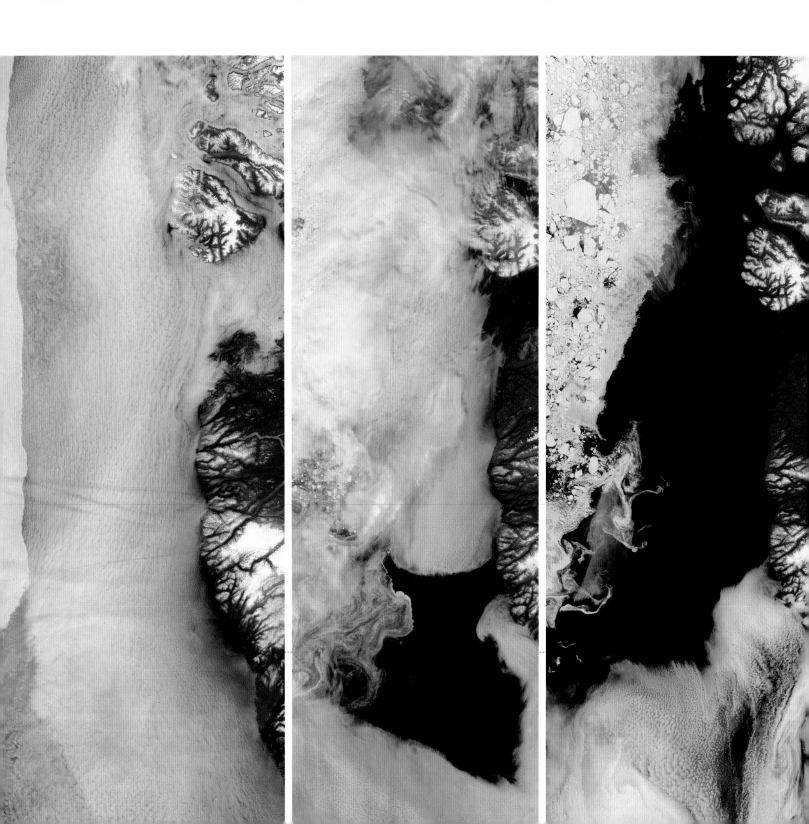

Climate change is the most important issue of the 21st century. The global economy—and civilization itself—may collapse unless greenhouse-gas emissions are controlled. Global warming already touches every part of the planet, and people everywhere are affected in their daily lives.

Changing weather patterns and the need to adapt to new conditions will dominate policy as the Earth's temperature and sea levels continue to rise.

Scientists believe that time is already very short—there may be as little as 10 years left to prevent irreversible climate change. Yet politicians, who have been made aware of the dangers, continue to act as though it were some far-off threat. As carbon dioxide builds up in the atmosphere at an ever faster rate, leaders of nations talk about the need for more talks to discuss how to solve the problem.

This book looks at the key issues. The science, the politics, what is happening in today's threatened world, and what (if anything) can be done. It is extraordinary how many people still do not understand the danger the planet is in. The information is available to anyone who cares to investigate, so there are some who believe that most of the developed world is in denial. They pick as an example the people who drive four-wheel-drive cars and take cheap flights to vacation homes, both of which use an enormous amount of fuel. They are the same people who, at the same time, make huge efforts to give their children the best education and start in life, assuming that life will be the same in 50 years. It won't be.

In Europe governments are worrying about pensions because of an aging population. They are encouraging 30-year-olds to start planning for their pensions. These young people, at least those who know about climate change and the economic havoc it will cause to the "safest" of our financial institutions, just laugh. They are almost certainly right to do so. For those who are in any doubt that climate change is real and happening now, this book explains the issues. It

explores other related environmental problems, which are made worse by climate change. Reducing poverty, improving water resources and sanitation, cutting air pollution, controlling tropical diseases, and saving species from extinction are all important. Add to that the need to increase food production to feed an ever-rising population on what will soon be significantly smaller landmasses, and you will see the difficulty.

The pictures in this book show the changes that are already obvious. Photographs from space document disappearing ice, forests, and the dust storms from enlarging deserts. To bring this into perspective, there are pictures of creatures in the wild that are already disappearing and may be extinct in our children's lifetime—or at least kept alive only in zoos, because their natural habitat has disappeared.

But while it may be sad to lose these and numerous other less majestic creatures, their fate is only an illustration of what may happen to much of the human race. Sea-level rise is going to wipe out the homes of millions of people. Islands in the Indian and Pacific oceans are already being evacuated because of rising waters. People are abandoning homes their forebears lived in for thousands of years. Villages are being made uninhabitable by saline intrusion into water supplies and high tides overtopping their islands. Other island countries, like the Maldives, have begun fortifying some islands as "safe havens" against the sea. How long will they last?

That is only the beginning. The small island states, the idyllic palm-fringed islands of the vacation brochures, are home to a few million people. But in Bangladesh 15 million live less than a meter above sea level, and in India there are another 8 million. These are the people with

Opposite: The amount of change between 2002 and 2004 on the west coast of Greenland, observed at the same time in June from an American satellite. Temperatures in Greenland and the Arctic generally have increased far faster than in the tropics, leading to more rapid ice melt than scientists had predicted and accelerating the rise in sea level. Scientists recently documented that the seasonal thaw along the margins of the Greenland ice sheet has been starting much earlier in the year and impacting a larger area than in the past century. As can be seen, sea ice in the region has been declining at the same time. More pronounced than the differences in completely snow-free areas are the increases in the total area caught in the act of melting—a translucent area sandwiched between the bare ground near the coast and the solid white of the interior of the island.

little money and no protection who will have to migrate to higher land. In an already crowded continent, it is hard to know where they can go.

But sea-level rise does not just affect the poor. Many of the world's largest and richest cities are at sea level. Dozens of cities will suffer total or partial inundation if they do not raise defenses against the sea. London and Rotterdam already have barriers that can be swung into place when high tides or storm surges threaten to overwhelm their defenses. St. Petersburg and Venice also realized some time ago that they are doomed without massive engineering works to keep the sea out. Fortunately for the people who live in these cities, they are on estuaries or lagoons where, with modern technology, barriers can be built across the entrances with gates to be closed when danger threatens. Others cities, including New York, Hong Kong, and Chittagong in Bangladesh are much more exposed.

Sea-level rise is just one of the perils the planet faces. Climate change makes a drastic difference to the food supply. The poorest continent, Africa, is already suffering from food shortages made far worse by a series of droughts, at least partly caused by climate change. In 2007 Australia was also enduring a five-year drought in an area that produces 40% of the country's agricultural produce. The Murray and Darling rivers, which provide 84% of the water used nationally for irrigation, are reduced to tiny streams, and some farmland has been abandoned. In all parts of the world, farmers are having to adapt to new temperature and rainfall patterns and grow different crops to suit the changed conditions. This mostly leads to lower productivity. More refugees already flee because of environmental factors than they do from war, according

Left: Floodwaters reflecting the grand façade of Belgrade's main train station on April 17, 2006, when the Danube broke through flood defenses. Thousands of people were forced from their homes in Serbia, Romania, and Bulgaria. The combination of rapid snowmelt in central European mountains and heavy rain caused this flood, one of a series in a five-year period when major rivers in Europe had burst their banks. Increasingly large areas of Europe can no longer obtain flood insurance because the number of these incidents has grown over the last 10 years.

On this page: Extreme weather events, such as floods, windstorms, and droughts, are increasingly common all over the world. *Top:* Rescue personnel search for people trapped in cars on a flooded street in Buenos Aires during a torrential overnight rain, January 31, 2004. *Bottom:* Flash floods have always been a feature of tropical regions, made worse by tree cutting and urban development. Vegetation soaks up rain, while hard surfaces turn roads into rivers. Warming oceans increase the violence of tropical storms. In October 2005 a Haitian man holds a woman in danger of being swept away during a storm in Port-au-Prince. The tropical storm Alpha brought heavy rain, flash floods, and mud slides to Haiti and the Dominican Republic.

"Stopping global warming is not just about saving the environment for the hunters, fishermen, hikers, and the other outdoor enthusiasts of today and tomorrow. Global warming is a matter of national security. Will we live in a world where we must fight our neighbors for fresh water and food?"

—General Wesley Clark

to the United Nations. In 2005 the news about climate change was bad. Global temperatures equaled those of 1998, the warmest year on record. It meant the unprecedented series of warm years from the last decade continues through this one. New data confirmed scientists' worst fears about existing warming trends and added some new and unexpected dangers. It is not just the addition of carbon dioxide from burning fossil fuels that is making the situation worse; the extra heat is believed to be releasing carbon locked in soil and methane in permafrost.

Parts of the Arctic were warmer than ever previously recorded, hastening the disappearance of sea ice and glaciers. Vast areas of permafrost, with millions of tons of trapped greenhouse gases, began to melt in Siberia. Svalbard (which is otherwise known as Spitsbergen), the group of islands well to the north of Norway, had the warmest winter on record. At the opposite pole, the West Antarctic ice shelves, thought to be stable, began to slide into the sea.

Scientists also discovered that the extra carbon dioxide in the atmosphere was making the oceans more acidic. For millions of years creatures have adapted to the existing conditions (for example, coral and shellfish relying on the alkaline seawater to extract the raw material for their shells). What effect will this new change bring about? The whole balance of the oceans may change. It has also become a lot clearer that sea temperatures are rising. Fishermen, who for centuries have hauled the same kinds of fish out of the same stretches of water, are finding that these fish have died out or migrated to find cooler places and been replaced by new, exotic varieties.

There are new fears that the ocean currents that transport heat around the globe are changing—and that the Gulf Stream might slow down or shut

off altogether, plunging parts of Europe into a much colder climate.

Partly because of increasing sea temperatures, 2005 was another year of extreme weather events. Over the last two decades these have steadily become more common the world over. Floods and droughts, which make headlines every year, were eclipsed because of the largest and most severe series of hurricanes ever recorded, made worse by the higher sea temperatures in the southern region of the North Atlantic and in the Gulf of Mexico.

Despite being the richest country in the world, with the best technology for predicting and studying hurricanes, the United States could not cope with Hurricane Katrina, which killed more than 1,300 people and devastated New Orleans. This is a city below sea level, prone to hurricanes, and aware of its vulnerability, yet it still failed to prepare for its fate. Despite this tragic lesson, the U.S. government is intent on spending billions of dollars rebuilding New Orleans in the same place. This is the sort of policy decision that leads some to believe that ours is a culture and a civilization in denial. This would not be the first. Civilizations have collapsed and disappeared in the past because of overconsumption. In most cases it has been overuse or misuse of a water supply. In others overhunting and overfishing wiped out the available food sources. In modern times Haiti is a classic example. This country of 8 million people was once mostly covered in forests, but only 2% remain because most were cut for firewood. The loss of forests meant the soil was washed away. The country is now suffering ecological and economic collapse, and only international aid is keeping it alive.

In every case the warning signs of impending disaster must have been there, but for some

"The identification of humans as the main driver of global warming helps us understand how and why our climate is changing, and it clearly defines the problem as one that is within our power to address."

—Union of Concerned Scientists, 2006

Top: Mongolian villagers migrate in search of ever decreasing grassland regions to feed their stock and escape from ever spreading deserts. Farther south, Chinese scientists report that there are "desert refugees" in three provinces: Inner Mongolia, Ningxia, and Gansu. The Asian Development Bank's preliminary assessment of Gansu province has identified 4,000 villages that may need to be abandoned as desertification intensifies.

Bottom: Thousands of internally displaced Kenyans wait in line at a water point on the road to the northern town of Wajir, 300 miles (500 km) from the capital, Nairobi, January 12, 2006. Kenya waived import duty on relief food to feed millions of people facing famine in the country's worst drought for years. The drought affected large parts of eastern and southern Africa, where the rains have become less reliable, which climate scientists predicted would happen as the climate warms. Partly as a result of the increasing difficulty of survival in the countryside people have moved into shanty towns on the edges of cities in the hope of making a living.

Above: A Chinese boy hugs his dog in front of his family home, set to be demolished after the water-level rise along the Yangtze River in Zigui, central China damaged its foundation. The water was rising with the filling of the mammoth Three Gorges dam reservoir. The area is higher than the water level of the reservoir and was supposed to be safe. China moved a total of 1.3 million people from areas affected by the controversial dam project, which is designed to provide hydro-electric power and control the flow of the river. The Chinese character reads "Demolish."

reason the political leaders failed to act in time. In the same way our civilization is steadily removing all its life-support systems at once, on a global scale, and at the same time changing the climate. It is not just events like those in New Orleans that lead to the conclusion that our political leaders are unable to face up to the crisis in civilization they are helping to create.

There is a point that needs to be made here about the responsibility of all of us, including scientists. Several times in researching this book I was told by some scientists that they could not be sure Hurricane Katrina or the melting ice caps was related to man-made climate change. It is what the computer models predict would happen and the sorts of events they would expect, but they cannot as yet be certain any one event is caused by the man-made greenhouse effect and is not just a natural phenomenon.

This is true, of course, but examine the evidence of all the remarkable weather events that occurred around the world in 2005, 2004, and the decade before, and it seems to be overwhelmingly obvious that something extraordinary was happening—in fact, that man-made climate change was here—a fact finally confirmed by scientists in the three 2007 Intergovernmental Panel on Climate Change reports. As well as those who, despite this, still claim they cannot yet see a definite link, other scientists, who accept that the enhanced greenhouse effect is already manifesting itself, say it is "up to us to show what is happening and up to politicians to decide what to do about it." It seems to me they are both passing the buck.

As a journalist who has been covering this issue for more than 20 years, it has been my training to state the facts, record the debate about whether climate change is real, and let readers make up their minds on the evidence

(continued on page 26)

Above: Bryn Mawr Glacier, Prince William Sound, Alaska, one of the many vast ice rivers that are now collapsing in the Arctic regions as temperatures rise faster than elsewhere. Glaciers are not just getting shorter, they are also getting thinner, releasing more and more water into the oceans and adding to sea-level rise. Only in Scandinavia, where precipitation has increased and heavier snow is falling in the mountains, are the glaciers apparently growing. Elsewhere across the mountain ranges of the world and near both poles, glaciers are shrinking, and in some places disappearing completely.

Left: Pancake ice is soft, newly formed sea ice that heralds the beginning of winter in the Ross Sea, the nearest open water to the South Pole. Within days of pancake ice appearing, the sea surface freezes solid, and any iceberg or ship caught in it will remain locked fast until the following spring, as the early explorers discovered. The sea ice is an essential breeding ground and protection for tiny sea creatures like krill, the main food sources of many species of whales and penguins. The extent of sea-ice loss each winter is leading to a reduction in the krill and a crash in the penguin population because they are not able to rear their young.

Right: Icebergs are formed when glaciers or ice shelves break off into the sea. Some are vast and take many years to break up and melt. They are a store of fresh water, and they change the salt content of seawater, which can in turn affect currents like the Gulf Stream. This is a natural process but it has speeded up in both the Arctic and Antarctic.

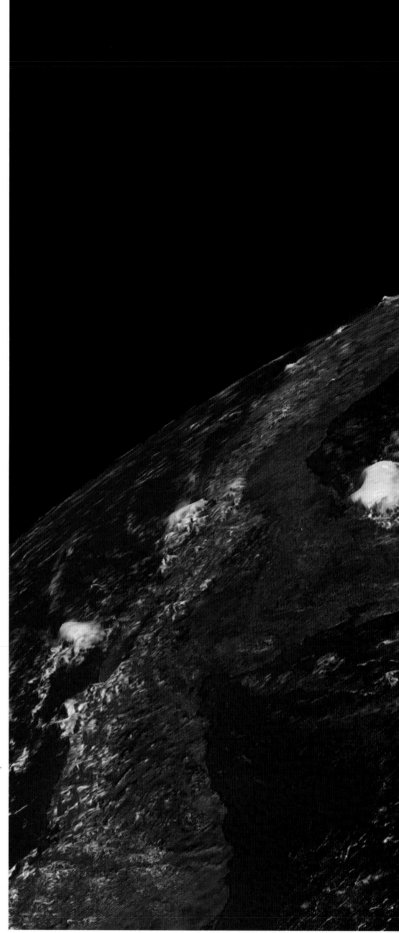

"After the tragedy of Hurricane Katrina, many Americans now believe that we have entered a period of consequences… they are beginning to demand that the administration open its eyes and look at the truth, no matter how inconvenient it might be for all of us—not least for the special interests that want us to ignore global warming.

"The climate crisis may at times appear to be happening slowly, but in fact it is a true planetary emergency. The voluminous evidence suggests strongly that, unless we act boldly and quickly to deal with the causes of global warming, our world will likely experience a string of catastrophes, including deadlier Hurricane Katrinas in both the Atlantic and Pacific."

—Al Gore in "The Moment of Truth," *Vanity Fair,* April 17, 2006

Right: The awesome sight of Hurricane Katrina, seen from space in this computer-enhanced image as it approaches the Mississippi Delta from the south-southeast on August 28, 2005, just before it struck New Orleans. The storm gained strength from a category 1 hurricane to a maximum category 5 as it crossed the Gulf of Mexico, where in 2005 the seawater was the warmest ever recorded, giving Katrina more energy to draw on. The low pressure in the center of a hurricane creates a dome of water in the ocean, which in this case overwhelmed coastal defenses causing serious flooding. Although New Orleans was warned days in advance and many citizens had escaped, many less fortunate people, without transportation, remained behind and were caught by the fierce winds, driving rain, and flooding when the inadequate levees collapsed under the weight of the water.

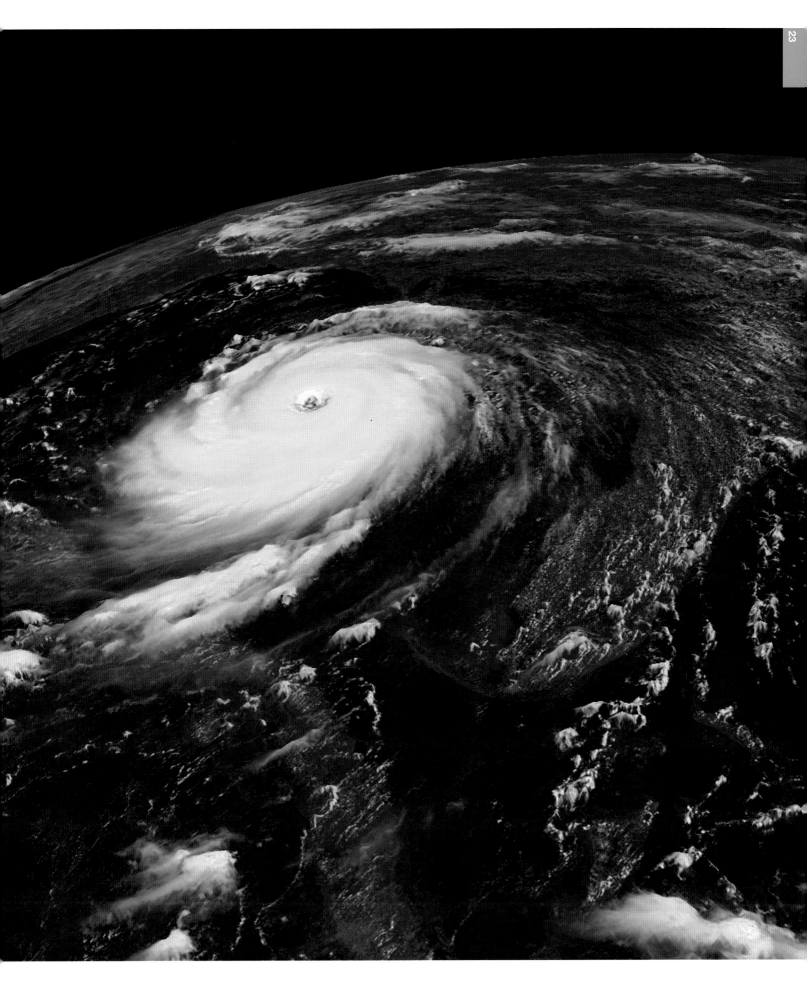

Above: The body of a victim of Hurricane Katrina lies in the remaining floodwaters and debris in the Saint Bernard area of New Orleans two weeks after the hurricane hit the city famous for its jazz. The images of bodies left uncollected amid the devastation of an affluent tourist city shocked the people of the richest country on Earth and the rest of the world. Because much of the city lies below sea level, the water could not drain away and had to be pumped out of the suburbs, delaying emergency services and evacuation.

Above: Stung by criticism of America's failure to cope with the damage caused by Hurricane Katrina and the administration's inability to look after the thousands of people left stranded without food and water, U.S. President George W. Bush toured Louisiana by helicopter. As he flew over New Orleans five days after the storm, he could see teams of rescuers still struggling to reach people in need of evacuation and looters raiding abandoned businesses and homes. Bush said it would take years to recover from the devastation experienced in Alabama, Mississippi, and Louisiana.

Opposite: The force of the hurricane and the giant waves whipped up by the winds damaged oil platforms, refineries, and installations all along the coast. These yachts were picked up by the floodwater from a marina near New Orleans—just a small part of the severe property damage along hundreds of miles of coast.

presented. That remains the correct approach for the news pages of a newspaper. For a scientist, that is also the right way to do things, at least when reporting new discoveries or theories in scientific journals. But surely both for scientists and the rest of us, responsibility does not stop there.

This is because as time has passed, it has become increasingly clear that the matter is urgent. The scientists had hoped that someone, somewhere, would do something—but this hope has not been realized. It is surely now time to take a step further. Scientists who are still hiding behind the uncertainties and the belief that they can do nothing more because it is someone else's job to act on the dangers they have revealed, should think again. So should the rest of us. As time has passed, I have come to the conclusion that waiting for someone else to act is no longer enough. Scientists and the rest of us have an ethical and moral dimension to our roles in society. We need to consider that time is short. Scientists are important because they are a group of people who have enough authority to wake the world up to the plight it faces. All of us need to play a part, and this book tells us how we can help.

Having that out of my system, it is fair to say that Europe, its peoples, and its political leaders are the most aware of climate change and its potential impacts. It is not clear why this is so; perhaps it is because environmental groups are most active, education levels are high, and the science base is so strong. Europe is credited with being the world leader in action on climate change. Tony Blair, then UK's prime minister, was at his most influential on the international stage in 2005. During his presidency of the international G8 forum, he made climate change a priority for the summit. To set the scene, he organized a conference to review the science. The news was

"The monumental ruins left behind by…past societies hold a romantic fascination for all of us. We marvel at them when as children we first learn of them through pictures. We feel drawn to their often spectacular and haunting beauty, and also to the mysteries that they pose. Lurking behind this romantic mystery is the nagging thought: might such a fate eventually befall our own wealthy society? Will tourists someday stare mystified at the rusting hulks of New York's skyscrapers, much as we stare today at the jungle-overgrown ruins of Maya cities?"

—Jared Diamond, *Collapse: How Societies Choose to Fail or Survive*, 2005

Above: Three civilizations that once had the time and resources to build great monuments—Easter Island in the Pacific (top), the Mayan city of Palenque in Mexico (center), and Ur in Sumeria in modern Iraq (bottom)—have faded away. History tells us that only when people have plenty to eat and have met all their basic needs do they have the leisure to construct symbols of their power and wealth. Why civilizations collapse is just as interesting, and the main cause seems to be overuse of scarce resources, exactly what is happening across the world in the 21st century. Archaeologists are still puzzling about why the Mayan civilization collapsed between A.D. 731 and 774, but in the case of Easter Island and Sumeria, the answers are known. The people of Easter Island cut down all of their forests and were unable to build the canoes that were needed to hunt for dolphins and other seafood, which was their staple diet. Both the population and statue cult collapsed. This Sumerian ziggurat at Ur in Iraq being overflown by a U.S. plane was abandoned 4,000 years ago when mismanagement of the irrigation system led to its soil becoming salty, food shortages, and the creation of a desert.

"Our generation has inherited an incredibly beautiful world from our parents.... We must not be the generation responsible for irreversibly damaging the environment."

—Richard Branson,
Global Initiative Summit in New York,
September 2006

shockingly bad. In response the political leaders made fine speeches but took no action to make any significant dent in the problem. Carbon dioxide emissions continued to rise alarmingly across the world.

It is true that in February 2005, after years of uncertainty, the Kyoto Protocol finally came into force and 34 countries accepted legally binding targets for reducing carbon dioxide emissions. Later in the year at Montreal, the world's nations met to finalize the details of the protocol and decide what to do to reduce emissions beyond 2012 when Kyoto expires. They decided to meet again in 2006 to talk about additional talks. One of those meetings took place in New Zealand in April 2006 when Tony Blair was again talking about the dangers of climate change. In a revealing moment he told the conference that asking for action on global warming from politicians was the "purest example" of the clash between "long-term interests and short-term gain." He said: "Often we are in a situation where it is not that governments do not want to do the right thing but that they worry electorally about the short-term consequences of doing so." This is well said, but it also shows clearly that our current crop of politicians are not statesmen.

So there is a large and ever greater divide between what the scientists are saying, clearly spelling out dangers for the future, and the politicians' feeble policy responses. The following pages explore some of the reasons why this has happened in the past. Of particular influence and interest is the way the fossil-fuel lobby has succeeded in looking after its own interests at the expense of the planet.

But all is not yet lost. There are signs that at all levels of society—in business, in communities, and through committed individuals—there are changes taking place. All over the world, city

(continued on page 30)

Above: Japanese children hold up a "planet of life" made of Shibori-zome, a silk tie-dye method traditional in Japan's ancient capital of Kyoto, on December 2, 1997. It was made for display during the city's hosting of a crucial global warming conference, which led to the Kyoto Protocol, the first legally binding treaty to provide targets for industrialized countries to reduce greenhouse-gas emissions by 2012. The "planet," a 2.7-yard-diameter balloon covered in the expensive fabric, expresses the children's hopes for blue skies and seas and rich land for a clean and beautiful Earth to be passed on to later generations.

"As for my own country, the Maldives, a mean sea-level rise of 2 meters would suffice to virtually submerge the entire country of 1,190 small islands, most of which barely rise over 2 meters above mean sea level. That would be the death of a nation."

—Maumoon Abdul Gayoom, president of the Republic of Maldives,
at the United Nations General Assembly, New York.
October 19, 1987

Left: Maumoon Abdul Gayoom, president of the Maldives, again pleads for his country as he addresses the Earth Summit Plus Five at the United Nations, June 24, 1997, on the perils of global warming for mankind. For more than 20 years the president has been warning world leaders of the plight of his nation and many other low-lying island states across the world. Leaders and envoys from 173 countries attended this special session of the General Assembly to review and appraise the implementation of policies adopted at the Rio Earth Summit in 1992, but few new actions were agreed upon.

Right, top: The Maldives is not just a tourist attraction. Like any nation, it has its own culture and traditional skills—in this case boat-building for the fishing industry. What will happen to this man and his craft skills when he has to move to another country and probably has to make a living miles from the sea?

Right, bottom: The Maldives looks like a wonderful place for children to grow up. But these children face an uncertain future because as sea levels rise, their homeland will disappear. By the time they reach middle age, the entire nation will be faced with evacuation to an as yet unknown land, and a loss of their culture and heritage.

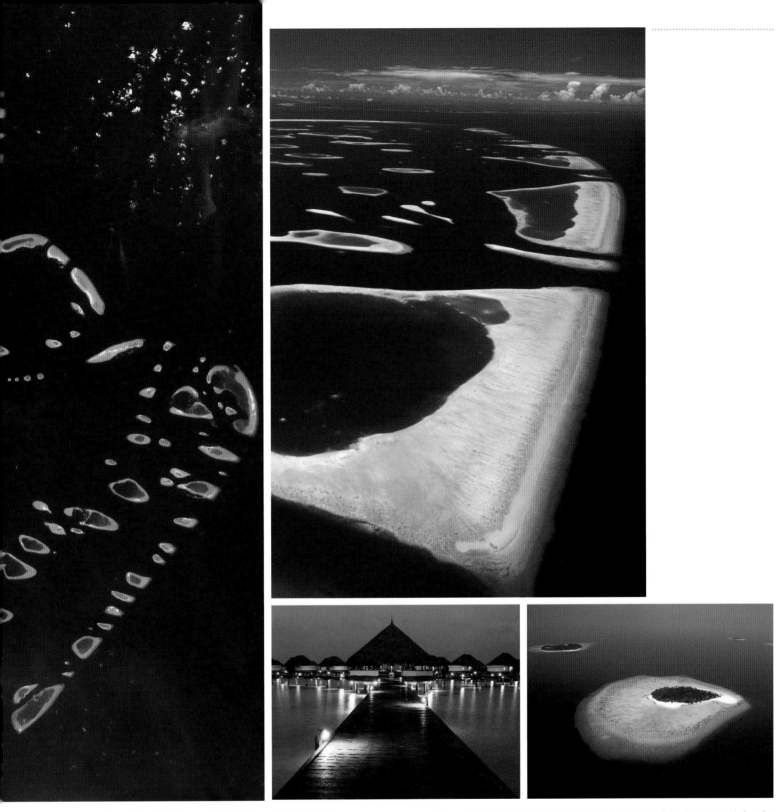

Left and above: North and South Malosmadulu Atolls, part of the Maldives, a republic in the northern Indian Ocean, southwest of India, with a population of 330,000 spread across a chain of 1,190 small coral islands with an average elevation of 1 meter above sea level. Waves triggered by the great tsunami of December 2004 spilled over the islands, forecasting what will happen as sea levels continue to rise. Some populated islands were so damaged that they were evacuated and others became mere sandbars in the sea. From space and aircraft it is possible to see how vulnerable the islands are, spread out on coral reefs in the Indian Ocean. At ground level the Maldives are still the idyllic islands that so many thousands of tourists visit each year without giving a second thought to the uncertain future that faces their dream vacation destination.

"Earth provides enough to satisfy every man's need, but not every man's greed."

—Mahatma Gandhi

mayors, local councils, and their voters are taking the threat of climate change seriously. Under the Kyoto Protocol, trading schemes designed to cut carbon emissions are under way. New clean technology is being transferred to people who would otherwise never have had the opportunity to develop without fossil fuel.

Carbon-trading schemes are now established across the world. The first was in Chicago in 2003, followed the same year by one in London, and in 2007 China became involved by opening an exchange in Shanghai. The idea is for various industries and countries that have caps on emissions to sell surplus carbon to others who cannot meet their targets. This provides incentives to industries, from farming to steelmaking, to become more efficient. The tons of carbon saved that would otherwise have been emitted to the atmosphere can be treated as an asset and this surplus traded on the world market.

The Europe-wide carbon-trading scheme, where every large company has a cap, should make a large impact on emissions. The first round failed to do so because governments, fearing competition, set their national caps on emissions too high, so industries could carry on polluting as before. This meant that the price of carbon slumped because hardly anyone needed to buy it. Lessons have been learned, however, and the price of carbon futures has risen to more realistic levels as caps have been tightened and companies find it more of a struggle to meet them and are forced to become more efficient or face a financial penalty in the form of having to buy carbon on the exchanges.

Across continents there are ingenious new ideas, inventions, and machines being developed that need to be made universally available. In other places old, well-proven ideas and methods, almost lost in the headlong rush toward a universal consumerist throwaway society, are being revisited and revised to suit new conditions.

Some of these heartening developments are reviewed in the following pages and their potential examined. As energy prices rise and oil begins to run out, as seems likely before the end of this decade, many of these technologies will be given a boost. New industries will boom as a variety of renewable technologies become heavily in demand to solve the looming energy crisis. This is good news both for society and the future of the planet. There is great potential for new jobs and export industries for those with vision, such as the wind turbine boom begun in Denmark. One of the great developments of this century will be small-scale energy production in the home, with wind, solar, and hydroelectric stations all grabbing a share of the market.

There is no single magic bullet that will solve the coming energy crisis, along with the danger to civilization and the natural world that climate change represents. But society and individuals can encourage change and create a political will for action by buying into these technologies. It requires individuals at all levels to take responsibility for their actions. At a personal level it matters what car you drive, what vacations you take, and even whether you use energy-saving lightbulbs (which we should all be doing already).

But individuals are also savers, shareholders, businessmen, bankers, architects, scientists, and politicians. Where do investors put their money, in clean or dirty industries? When consumers buy products, such as wood or fish, they must demand to know that they come from sustainable sources. Scientists can no longer continue to hide in their ivory towers. If it is too much to hope that politicians can be persuaded to care even if the next election does

Opposite: Every major city in the world, rich or poor, now boasts traffic jams. Shown here is New York's Times Square, choked with traffic on November 30, 2004, and Dhaka, the capital of Bangladesh, at roughly the same time. For both cities air pollution at street level makes for an unhealthy life for both pedestrians and motorists. Before 2002, Dhaka laid claim to be the most-polluted city in the world. More than 50,000 two-stroke three-wheel taxis or rickshaws, known locally as baby taxis, were spewing black toxic smoke onto the streets. Levels of pollutants reached six times World Health Organization maximum levels for lead, carbon monoxide, and deadly particles that lodge in the lungs. A government ban on two-stroke vehicles in 2002, replacing them with four-stroke "green" taxis that run on compressed natural gas, has improved the situation, but traffic congestion and pollution are still chronic.

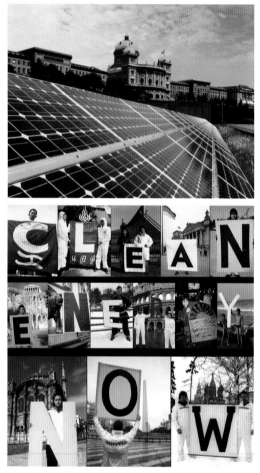

not depend on it, can voters persuade them that green policies count at the ballot box, too? Writing to your congressional representatives and local newspapers about your concerns is always a good start.

Perhaps what is most surprising and encouraging about the battle to save the planet from man-made climate change is that there are people at every level of society committed to the cause. And this is not just in the old industrial world, where wealth has already been created at the expense of the environment.

The accepted wisdom in the United States and most of Europe is that in countries like India and China, the chief concern is economic development; environment comes a distant second place. While this has been partly true, both countries realize the terrible cost in the health of their people and future prosperity climate change will bring. These countries, the most populous in the world, now want clean development and are beginning to make it a priority.

China, in particular, has made a policy decision to try and curb growth to make it more sustain-able. Following massive dust storms, record vehicle pollution, and grit from building sites that choked Beijing in April 2006, the prime minister, Wen Jiabao, ordered a reconsideration of the situation. The 30% increase in hospital admissions because of smog, and the warnings that children be kept indoors for their own safety, led the prime minister to say that the policy of economic growth at all costs had been wrong. He said measures to conserve nature and improve air and water quality over the previous five years had not been successful, and had resulted in severe ecological degradation.

The government had failed to reach eight of its 20 environmental goals. From 2006, every six

months local government would be obliged to release information on energy consumption, which must be cut by 20% by 2011, and on the emissions of polluting chemicals, which must be reduced by 10% over the same period. The country needed to focus on the consequences of growth, he said. In 2007 it was clear that the government's change of emphasis was so far having little effect. The Chinese economy was reported to be growing even faster at more than 10% a year.

In India, the government is active in reducing air pollution in some cities. In New Delhi, the capital, where 70% of the pollution came from traffic and the levels of dangerous particles in the air were 10 times World Health Organization limits, the supreme court ruled in 2003 that the entire public transportation fleet must run on clean fuel to protect the population. As a result 80,000 vehicles, including 9,000 buses and 45,000 auto-rickshaws, have been converted to run on natural gas. There has been a dramatic reduction in carbon dioxide and other polluting emissions, although the growth of private vehicles threatens to wipe out the improvement.

Across the world, the detrimental economic consequences of climate change are at last being understood, and it is not just the environmental lobby or scientists saying so. On October 30, 2006, Sir Nicholas Stern, former chief economist at the World Bank, delivered a 700-page report to the British government in which he said the world economy would shrink by 20% unless climate change was tackled urgently. Just as important, he said that averting the crisis would only cost 1% of the world's income if leaders acted now. He said if action were not taken, then in the next 50 years, floods from rising sea levels would displace up to 100 million people, melting glaciers could cause water shortages for 1 in 6 of the world's population, up to 40% of wild

Top: Switzerland is in the front line of climate-change effects, with winter warming in the Alps damaging its vital skiing industry. The country has reacted by investing in alternative energy, as this photovoltaic electricity production plant on top of an office building in Bern shows.

Bottom: The G8 group of industrialized countries, which so far have talked a lot and done little about climate change, confronted by Greenpeace volunteers in Trieste, Italy, in 2001. Fourteen activists, each from a different country, demand clean energy.

Opposite: Middlegrunden, off Copenhagen in Denmark, became the world's biggest offshore wind farm in 2001 with this impressive curve of turbines stretching 2 miles (3.4 km) across the shallow sea not far from the city. The 20 2-megawatt turbines provide 3% of the Danish capital's electricity. Since this wind farm was built, turbines have increased in size and are soon each expected to have an output of 5 megawatts. The pioneering Danes are exporting the technology, supporting 20,000 jobs, and creating one of the country's largest and most profitable industries.

"How much longer must we sit by and watch an increasing number of unusual natural disasters unfold before we realize that something is amiss? The small island countries warned more than a dozen years ago that there would be an increased frequency and intensity of storms and other adverse weather events as a consequence of global warming. In a sense they have served as mankind's early warning system...the canaries in the cages in the coal mines. Is anyone listening?"

—Robert Van Lierop, writing in 2004 for the Institute of the Black World. He was one of the pioneers in 1990 negotiating the text of the Climate Change Convention.

species could become extinct, and droughts would create tens or even hundreds of millions of "climate refugees."

Some commentators thought that even though the predictions of the Stern report were dire, it did not go far enough. This was because the findings were based on the 2001 predictions of the Intergovernmental Panel on Climate Change. Since that report was published, as became clear in the undated 2007 version, climate change has speeded up considerably and its predicted economic disadvantages have become much clearer and closer at hand.

Nevertheless, the report was heralded by environmentalists as incontrovertible evidence of the need for immediate action and said that continuing with business as usual was the "economics of genocide." Politicians were also universally quoted as both accepting the findings of the report and the fact that immediate action was necessary. Four months later then UK Chancellor, Gordon Brown, who had originally commissioned the Stern Report and immediately accepted its findings, tinkered with a few green taxes in his budget. He raised the road tax on expensive gas guzzling cars, fuel taxes for vehicles, and doubled the small airport taxes, but these measures were not expected to make the slightest impact on pollution levels, the growth of air travel, or change people's behavior in any way.

What was interesting about the Stern report and about most modern environmental thinking is that there is nothing about the future that is envisaged that implies a reduced quality of life; in fact, the opposite is true. The idea is that rapid adjustments to new technologies will allow both the existing generation and the next to enjoy this rich world, rather than watch helplessly as runaway climate change takes hold. It does not

mean that some older industries and the people who work for them will not suffer changes as newer technologies grow to replace them. But these technologies will provide cleaner alternative methods of transportation and less polluting electricity and heating plants. This is a process that is happening anyway and would be forced on all of us eventually as fossil fuels run out.

As we can see in the final chapters of this book, communities across continents are already making painless changes to wean themselves off fossil fuels for cooking, heating, and lighting, and providing themselves with a clean water supply and sanitation. As they do so, they improve their own quality of life. In the manufacturing sector the race is on to exploit clean technologies and create new industries and thousands of jobs. In older industries, car manufacturing, for instance, competition from Japan is forcing U.S. manufacturers to rethink their range of models. The competition across the world is focused on finding ways to mass-produce vehicles that are not solely dependent on oil for propulsion. But much more needs to be done, and done quickly.

With international political leadership still going far too slowly to solve the problem, it is up to committed individuals at city and community levels to make the difference. Increasingly it is becoming clearer that personal choices need to be made about how we can each lead our lives to reduce our carbon footprint. This is not always easy to do while the world is still hell bent on an unsustainable path, but every day new products and new opportunities appear for individuals to help change the world.

The size of the problem is frightening, but there is still time, just.

Opposite, top: Deserts can bury villages as the sand dunes move across the country like waves on the incoming tide. In Mauritania the women of Ljnanoune drag a giant net to the top of the encroaching dunes to try and hold back the Sahara Desert from smothering their homes and fields.

Opposite, bottom: The rapid retreat of the Gurschen Glacier in the Andermatt region of Switzerland led the ski resort to cover ice with a specially made fleece at the start of the 2005 summer in a bid to cut the rate of melting by blocking out the sun's ultraviolet rays. The shrinking of the glacier has meant that the resort, which attracts 250,000 visitors annually, has had to build a larger ramp each year to get skiers to the slopes. The 4,374-sq.-yd. (4,000-sq-m) fleece is designed to cut ice loss by 75%. The alpine glaciers are losing on average 1% of their mass each year and threatening many of the ski resorts with bankruptcy.

The Canadian Arctic is warming faster than many scientists predicted. This climate change is melting Arctic sea ice at an alarming pace and presenting serious implications for the people and wildlife that live there. Polar bears, for example, who swim in near freezing water to hunt for food, are literally drowning from the lack of ice cover.

The Arctic is warming twice as fast as the rest of the planet; 2005 saw Arctic ice cover at its lowest for over a century. This came after four consecutive years of reduced ice cover, a circumstance that experts conclude can only be the effect of climate change.

Drought-related forest fires during the 2005 season in the United States ravaged over 8.6 million acres, causing close to a billion dollars' worth of damage. States most affected in recent times include California, New Mexico, and Arizona.

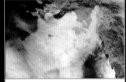

The European heat wave of 2003 is said to have been responsible for over 49,000 deaths, mostly among the elderly, sick, and the very young. Throughout August, Europe experienced record temperatures (August 10 was the hottest day in the UK since records began), forest fires, and flash floods due to intense thunderstorms.

The growing scarcity of oil has led to enormous conflict between exporters and importers. Although different reasons were given for the conflict, the United States, the world's largest importer, led an invasion into oil-rich Iraq in 2003, which has led to the deaths of thousands of coalition servicemen and women and tens of thousands of Iraqi men, women, and children.

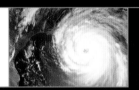

In 2005, the Gulf of Mexico saw the worst hurricane season since records began with 14 hurricanes. Hurricane Katrina was the most devastating, causing up to $26 billion of damage and claiming more than 1,000 lives. Extreme events like these will be a feature of a planet with warming seas.

Logging, road development, cattle ranching, soya cultivation, and colonization are among the main culprits for deforestation in the Amazon. In 2003-2004 Brazil lost more than 10,039 sq. miles (26,000 sq km) of forest, an area over 16 times the size of London. The area suffered a severe drought in 2005.

Everywhere you go, the effects of climate change are evident. It is not a problem limited to poor or remote areas, but a problem that threatens all of us. As this map shows, a warming world will cause extreme weather events, health crises, and the destruction of some of nature's most wondrous treasures. We cannot afford to keep ignoring this problem. It is the last chance for change.

Experiments on the Antarctic ice core show that human activities have increased atmospheric greenhouse gases by an unprecedented rate and volume. The Antarctic itself is suffering from glaciers in retreat as a result of climate change, destroying the habitat of many of its unique species.

The Danube floods in 2006 saw record levels of floodwaters and the Balkan countries Romania, Bulgaria, and Serbia facing huge numbers of evacuees. Melting snow combined with heavy spring rain caused the river to overflow its banks, reaching its highest level for over a century, 26 ft. (8 m).

In Siberia permafrost, the so-called permanent layer of frozen soil, which has existed since the last Ice Age, is melting. An area of a million square km is being transformed into a mass of shallow lakes and could unleash up to 70 billion tons of methane into the atmosphere.

In China advancing deserts are causing dust storms that stretch as far as the western United States. Land lost to desert each year has increased from 602 sq. miles (1,560 sq km) between 1950-1975 to 1,389 sq. miles (3,600 sq km) at the end of the 20th century. The Gobi desert is now within 93 miles (150 km) of Beijing.

The U.N. has estimated that approximately 11 million people are in serious danger as a result of drought in East Africa in 2006. Christian Aid has gone further to suggest that climate change could be responsible for the deaths of 184 million people in Africa alone in the years to come.

Bangladesh has a population of close to 148 million people, a population density that averages just over 1,100 people per sq km. The floods that have long ravaged this land have become worse in recent times. This, plus sea-level rise, threatens a third of the land area.

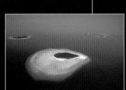

The Maldives, in the Indian Ocean, is an ideal paradise that, as a result of rising seas, faces total inundation. As Maldivian president Maumoon Abdul Gayoom warned the U.N. at the General Assembly in 1987, this would signify the "death of a nation."

Coral bleaching is taking its toll on one of the world's most breathtaking natural phenomena, the Great Barrier Reef. Rising sea temperatures and changes to the chemistry of the ocean have caused widespread bleaching and coral disease, leading experts to believe that there will be no living coral in the Great Barrier Reef by 2050.

How Close Is Runaway Global Warming?

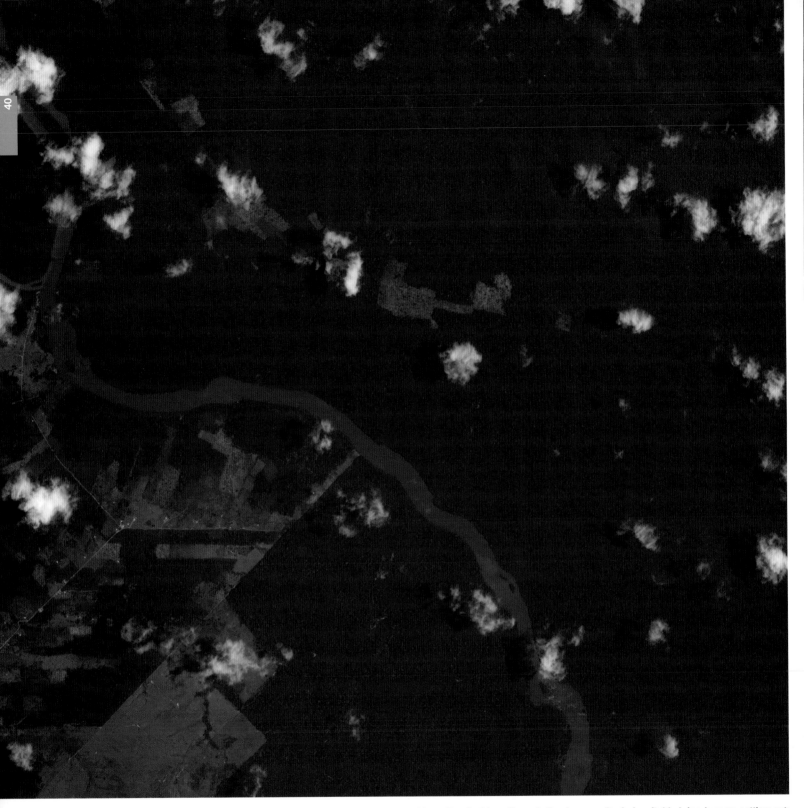

Previous spread: Teenagers run from waves breaking over sea barriers in the northern Taiwanese city of Keelung on October 16, 2001. It was a dangerous adventure. This typhoon, Haiyan, was the ninth storm to hit the island of Taiwan that year. The violence of the typhoons broke records for the amount of damage caused to Taiwan and the number of lives lost. All over the world the violence of storms is increasing, which is in line with the predictions of scientists using sophisticated computer models to look at future weather patterns.

Above: Roads driven through the Amazon Basin inevitably bring human settlement and development. This high-resolution satellite image of land on either side of an Amazon tributary shows farms cleared from the jungle spreading on either side of several new tracks. Where there are no roads to the north of the river, the jungle is virtually unbroken. The river itself is colored brown by sediment, possibly from deforestation upstream. The picture is part of a scientific study of the region to see the effects of human-caused change in the region and the possible effects on climate.

Above: The stark contrast between virgin jungle and land cleared from farming in the Mato Grosso state of Brazil, one of the greatest areas of forest destruction in the Amazon Basin. The soil is poor, and once cleared of trees, it rapidly loses its moisture and fertility and in some areas has already turned into semidesert. Attempts by the Brazilian government to stem the tide of destruction have so far failed, because many of the frontier regions of this vast country remain lawless and corrupt.

The slow political reaction to the increasingly dire warnings of scientists about the fate of the Earth has left many who have been following events shaking their heads both in sadness and in disbelief.

The European Union, more prepared than any other power bloc to face the facts, has accepted scientific advice that any more than a 2°C rise in temperature above preindustrial levels risks catastrophe. Above that temperature natural systems will find it hard to adapt. Many specialized species will be unable to thrive in a changed climate, and complex relationships between plants, insects, and animals will be thrown into chaos.

But worse than that, an increasing number of scientists believe that above that temperature, climate change might get out of control, at least beyond the ability of mankind to control it. There are two different ideas here: runaway climate change and a so-called tipping point. The first term sounds a bit like science fiction, but this idea of runaway climate change is not new. It is what has happened on Venus, our nearest planet, where the atmosphere is untenable for life as we know it on Earth.

Runaway climate change is a theory of how things might go badly wrong for the planet if a relatively small warming of the Earth upsets the normal checks and balances that keep the climate in equilibrium. As the atmosphere heats up, more greenhouse gases are released from the soil and seas. Plants and trees that take carbon dioxide out of the atmosphere die back, creating a vicious circle as the climate gets hotter and hotter.

The second phrase, tipping point, is heard a lot more from scientists. That is where a small amount of warming sets off unstoppable changes—for example, the melting of the ice caps. Once the temperature rises a certain amount, then all the ice caps will melt. The tipping point in many scientists' view is the 2°C rise that the EU has adopted as the maximum

limit that mankind can risk. Beyond that, as unwelcome changes in the Earth's reaction to extra warmth continue, it is theoretically possible to trigger runaway climate change, making the Earth's atmosphere so different that most life would be threatened.

As with much of climate science, what used to be theory is now being seen in practice on the ground. New information makes clear that reaching the tipping point is a much more immediate threat than was previously thought. Some of the causes of the deep concern that we might lose control of our own destiny are discussed below. The danger grows with the increase in average temperature above what is called the preindustrial level—the mid-18th century. Some scientists estimate that when the temperature reaches an extra 2°C above that equilibrium, the Earth's natural systems will be in serious trouble. It will affect many species' survival prospects, including our own.

So the key question is: How close are we to a 2°C rise, and when will we get there? The first thing to admit is that nobody knows for sure, but many who understand the science say the answer to this twin question is, first, that we are already very close, and second, we might get there terrifyingly soon. In fact, the 2°C threshold is much closer than almost anyone outside the specialist scientific community is prepared to acknowledge.

By any standard, if you care about the future of the human race, it is too close for comfort. Warming is directly related to the quantities of greenhouse gases there are in the air, the chief of which is carbon dioxide.

Concentrations of carbon dioxide in the atmosphere were already at 383 parts per

Above: A forest fire arriving in the town of Malhao in the southern Portuguese province of Algarve in 2004. It is one of dozens of fires that ravaged the countryside during droughts and heat waves over a three-year period. This resident was unable to save his home and ran for his life. Hundreds of firefighters supported by planes and helicopters, some sent from France and northern Europe, battled against the forest fires across Portugal in a heat wave that reached a high of 104ºF (40ºC). Portugal had already lost 13% of its forests and woodlands in 2003, before these fires caused further destruction.

"One person flying in an aeroplane for one hour is responsible for the same greenhouse gas emissions as a typical Bangladeshi in a whole year."

—Beatrice Schell, European Federation for Transport and Environment, November 2001

Sources of Greenhouse Gases

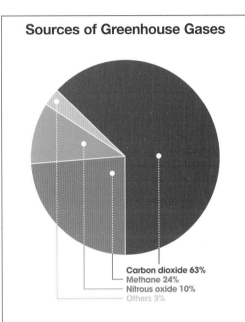

Carbon dioxide 63%
Methane 24%
Nitrous oxide 10%
Others 3%

Although there are a number of greenhouse gases, carbon dioxide is the major contributor to warming expected over the next century. It is the burning of fossil fuels that is directly responsible for the CO_2 increase in the atmosphere. It takes approximately 100 years for carbon dioxide to be absorbed back into living plants. Methane disappears more rapidly, but it still needs urgent controls because it is a more potent warming gas that comes from agriculture, rotting garbage dumps, oil and coal exploration, and leaky pipelines.

SOURCE: Hadley Centre for Climate Prediction and Research.

million (ppm) in 2007. That is up from the pre-industrial level of 280 ppm, a considerable increase. To put that in perspective, we need to realize that the 280 ppm figure had remained more or less unchanged for 10,000 years—the period that accounts for the entire span of modern human history. The benign climate that has allowed the human race to multiply, develop, and prosper has remained stable through that period. There have been minor variations: warm periods that allowed places like Greenland to be settled by the Vikings or medieval monks to make wine in Britain, and cold periods, known as mini ice ages, that made it possible to have frost fairs on the frozen Thames in London during the 17th and 18th centuries. The last one was held in the winter of 1814.

These so-called natural variations in the climate have given plenty of ammunition to those trying to discredit global warming theories. But those changes have now been well studied and are better understood. It is no longer credible to suggest that what is happening now is a natural variation of a sort recorded in the last 2,000 years. In fact, the variations in the quantities of carbon dioxide in the atmosphere have been small in that period, and other natural variations, like sunspots, have been the culprits for the previous warm and cool periods. The recent increases in greenhouse gases have changed all the rules and the stability in the climate system man has enjoyed for so long.

Current calculations suggest that if and when the level reaches 450 ppm, there will be a 50% chance of the Earth's temperature exceeding a rise of 2°C—in other words, an even chance of potentially catastrophic climate change. To be on the safe side (the so-called precautionary principle, which so many politicians claim they

endorse), some scientists believe that the carbon dioxide in the atmosphere must be pegged back to 400 ppm—a mere 17 ppm above the current level. So on their current calculations, since man began the Industrial Revolution and unwittingly an experiment with the climate, the human race is already more than 80% on the way to causing a potential disaster.

With this evidence it is clear that drastic action is needed. Some scientists have certainly been urging politicians to take immediate measures. Recent evidence demands, according to a consensus of the world's best climate scientists, that we need to cut existing emissions by between 60% and 80% in the next 40 years to stand a chance of preventing climate change from becoming unstoppable to keep control of our own destiny. Compare that figure with that achieved by the Kyoto Protocol—to date the best effort by politicians to cut emissions. This will cut greenhouse gases from 34 of the developed countries by 5.2%, excluding the United States, the world's biggest polluter. Over the period of the agreement, which lasts only until 2012, total world emissions will rise because of the growing industries of the developing world.

What does science tell us about how much time we have left to solve the problem? Measurements taken by NASA's Goddard Institute for Space Studies and the Earth Institute of Columbia University, released in December 2005, show that in the last 100 years the world's average temperature has increased by 0.8°C. That seems to leave a comfortable 1.2°C to go before the tipping point is reached, but this is where the climate plays a nasty trick. Unlike glass in a greenhouse, the extra heat-trapping gases released into the air take time to build up to their full effect. That is largely because of

Opposite: The tripling of the price of oil has made mining oil sands a lucrative business. The largest deposits in the world are in Northern Alberta, Canada, in an area of 57,000 sq. miles (149,000 sq. km), making it the world's second largest oil deposit after Saudi Arabia. Over 82,000 acres (33,000 hectares) of pristine forest have already been destroyed to reach the oil in a rapidly growing industry to the dismay of environmentalists. The resident wildlife is being wiped out and the Athabasca River, the main source of the high volume of water needed in the mining process, is being polluted and depleted. Continued mining will make it impossible for Canada to reach its Kyoto greenhouse gas reduction target and will turn the country into an environmental pariah state.

the delaying effect of the cool oceans as they catch up with the atmosphere. Best estimates are that there is a 25- to 30-year time lag between the greenhouse gases being released into the atmosphere and their full heat-trapping potential taking effect. That wipes out any feeling of comfort. It means that most of the increase of 0.8°C seen so far is not caused by current levels of carbon dioxide but by those that already have been in the atmosphere up to the end of the 1970s. Even worse, the last three decades have seen the levels of greenhouse gases increase dramatically. In this 30-year period the Earth has seen the largest increase in industrial activity and traffic in history. This great burning of fossil fuels has also coincided with the mass destruction of rain forests. So on top of the extra heat we are already experiencing, there are another 30 years of accelerating warming built into the climate system.

Using their best understanding of how the climate works and the largest computers in the world to create a picture of what is happening, scientists believe that in those three decades enough extra gas has been released to raise the Earth's temperature by as much as 0.7°C on top of the increases already measured. Add the 0.8°C already measured to the 0.7°C, because of the last 30 years of growing emissions, and the human race is already committed to a total rise of 1.5°C.

The full impact of this needs further explanation. As it has become increasingly clear over the last 15 years, the more you know, the worse the prospects appear to be. In this case it is the annual increases of carbon dioxide in the air. When annual measurements began 50 years ago, the average rise a year was around 1 ppm, sometimes dipping below, often slightly more. By the end of the 1990s, the average increase

had risen to 1.5 ppm. In the first five years of this century, it has exceeded 2 ppm twice (2.08 ppm in 2002 and 2.54 ppm in 2003) before dropping back to 1.5 ppm. In 2005 it rose more than 2 ppm again and in 2006 another 2.6 ppm. Over 50 years the amount of carbon dioxide trapping heat in the air has been on a general upward curve and now appears to be accelerating.

So in terms of the potential for runaway climate change, what does all this mean? A paper presented to the International Symposium on Stabilization of Greenhouse Gases in the Atmosphere in Exeter, England, in February 2005 gave a stark warning. Malte Meinshausen of the Swiss Federal Institute of Technology said that to be "on the safe side"—that is, not to gamble with the future of the planet—it would be sensible not to exceed 400 ppm of carbon dioxide in the atmosphere. At the beginning of 2007, the level had already reached 383 ppm. At a minimum this figure is rising by an average exceeding 1.5 ppm a year, so unless there is a drastic change in human behavior or some other unknown factor, the 400 ppm safety threshold will be exceeded within 10 years.

And this deals only with carbon dioxide. This is the main greenhouse gas, and the most important, but others also make a significant contribution to warming the atmosphere, particularly methane and nitrous oxide. Levels of both of these are also rising above preindustrial levels. According to the World Meteorological Organization, methane was at 1,783 parts per billion (ppb) in 2004, and nitrous oxide was at 318.6 ppb, the highest ever recorded. This is an increase on preindustrial levels of 155% and 18%, respectively, and an increase in a decade of 37 ppb and 8 ppb in absolute amounts.

Left: A dust storm in Beijing turns the sky an amber color and reduces visibility to 587 yards (500 m). This picture was taken in March 2002. Dust storms are a regular feature of life in the capital of China as the deserts creep ever closer to the capital, settling only about 100 miles (150 km) away. Extensive deforestation and desertification in Northern China have fueled the dust storms, and the Chinese estimate that nearly 1 million tons of Gobi Desert sand blows into Beijing each year. The authorities are hoping this will not happen during the 2008 Olympic Games in the city.

"We are upsetting the atmosphere upon which all life depends. In the late '80s when I began to take climate change seriously, we referred to global warming as a 'slow motion catastrophe,' one we expected to kick in perhaps generations later. Instead, the signs of change have accelerated alarmingly."

—Dr. David Suzuki, founder of the David Suzuki Foundation, October 2005

Scientists calculate that by adding the warming effect of these to carbon dioxide, the equivalent warming effect is that of another 45 ppm of CO_2 in the atmosphere. If this is correct, then the world is already well over the 400 ppm level, so it is almost impossible to avoid dangerous consequences.

Some scientists think this is too pessimistic a picture, because the human race can also adapt to the changes that climate change will bring. Professor Sir David King, the UK government's chief scientific adviser, giving a talk to senior industrial figures at Reuters in London in 2006, said the current target was to keep below 550 ppm. Afterward I challenged him about the figure; surely, that was taking a gamble and risking dangerous climate change, wasn't it? He conceded that 400 ppm was the "scientific ideal" but that politically it was not attainable. It was no use asking politicians to go for a figure they could not reach.

But if the figures given at the Exeter conference are right, then Sir David was certainly advising politicians to take a serious gamble with the future of the human race. In a subsequent interview with the BBC, he clarified his position. Current British government policy was to cut carbon dioxide emissions by 60% by 2050, but this was based on keeping below 550 ppm in the atmosphere. He conceded that based on the latest science, that would take the temperature up 3°C by the end of this century, well above the "danger" threshold. His solution was to go for adaptation—building sea walls, moving cities, and changing food crops. "After all, we have nearly 100 years," he said. He conceded, however, that the developing world would be hard hit by this approach and that some countries, particularly small island states like the Maldives, would disappear altogether. The

latest report to the Hadley Center for Climate Prediction and Research in Exeter said a 3°C temperature rise would cause a drop worldwide of between 20 million and 400 million tons in cereal crops and put about 400 million people at risk of hunger.

Despite these dire forecasts, some optimists argue that even if the 400 ppm threshold is exceeded, it would still be possible to reduce current emissions over the next 50 years or so to bring the levels back down toward 400 ppm or even below. That would give the atmosphere a chance to stabilize and keep the ice caps from melting. The kind of effort required to achieve this would be enormous. In fact, huge strides in technology and political effort would then be required to get the 60% to 80% reduction in man-made emissions over the next 50 years. This is, as previously said, the level of cuts scientists believe is the minimum required if we are to avoid "dangerous climate change."

There is more about the issue of dangerous climate change in the next chapter because it is an important phrase in the political sense. All of the governments that have signed up for the 1992 Climate Change Convention, and there are 180 of them, have signed up to take measures to avoid "dangerous climate change." So far they appear to have avoided asking themselves what this actually means in practice. But on any calculation, whether politicians accept 400 ppm as a target or even the much laxer and riskier 550 ppm, time for action is short. At best, we probably have only until 2020 to make decisions and change the way we power the planet to give the human race a reasonable chance of avoiding dangerous climate change. Based on current progress, that is a tall order. At first sight it is hard to understand why this potentially imminent economic and social

Opposite, top: A Kenyan man walks with his donkey carrying water after trekking 4 miles (6 km) to the only well with water in El Wak, 950 miles (1,530 km) from Kenya's capital of Nairobi, in December 2005. The Mandera district was littered with carcasses and crisscrossed by nomads hunting for water as the remote plains of northern Kenya suffered a drought that killed hundreds of thousands of livestock. The area, which borders Somalia and Ethiopia, was already one of Kenya's poorest and most arid areas before the drought struck.

Opposite, bottom: In another continent the desert is also still advancing despite efforts to stop it. A Chinese primary school student surveys barren fields after taking part in an afforestation project at Huai Lai county in Hebei province, 45 miles (70 km) northwest of Beijing. The environment in Huai Lai county has been deteriorating so much that parts of fields and several small mountains have been buried in sand as a result of desertification. Official statistics show that over the past decades, more than 40% of China's territory has been turned into desert, contributing to the dust storms that can reach as far as the United States. Before the 1980s, China's desertified land expanded at an average rate of 625 sq. miles a year (1,560 sq km). The figure rose to 840 sq. miles a year (2,100 sq km) in the 1980s and to 980 sq. miles (2,460 sq km) in 1994.

catastrophe is not discussed more. The science is not difficult to comprehend; the average 10-year-old could grasp the basics of it. Yet the consequences are so shocking that the standard response from those in power is to avoid the issue of how close we are to this dangerous threshold.

Cynical observers might conclude that politicians, while in office, seem to fear that if they dwell on such a question, the consequences of the answer might limit their own survival. At first sight, at any rate, they will have to do something unpopular, something the public won't like. So they fear that if they protect the future of mankind, the current generation will resent it and vote them out of office. Although heads of state like to be called leaders, they appear to believe that voters, like the much maligned lemming, would rather run over a cliff than be told they have to stop running in order to avoid mass suicide. Experience of what is happening around the world, and there are many examples, show this is not so. Given the opportunity, the correct information, and access to the already proven technologies, both individuals and communities are prepared to work hard to combat climate change. It is also clear that if everyone were given the chance to be involved, the problem could be solved.

The collective failure of the heads of the world's most important countries to lead their populations away from the abyss is examined in later chapters, along with ways in which the situation could be remedied. But to comprehend the extent of the leaders' dereliction of duty, it is important at this stage for the reader to understand the science, which all leaders have been made aware of but would rather pretend they had not been.

Left: Forest fires in the hills above the suburbs of Los Angeles on October 25, 2003, as seen from a satellite. Huge wildfires are burning to the east of the city. Just one day later the situation was even worse, with several massive fires raging across the region, driven by the fierce Santa Ana winds that blow toward the coast from the interior deserts.

Above: Los Angeles was one of the first cities in the world to realize that traffic
was causing a major pollution problem when it became famous in the 1970s
for its smog, caused by the action of sunlight on exhaust fumes. This aerial
fish-eye–lens view of Los Angeles at twilight shows part of the vast area the city
covers and its network of highways. The problem of smog continues despite the
introduction of catalytic converters, and it is now compounded by the threat
of global warming. In response the state of California has become a pioneer in
encouraging low-emission vehicles and progressively tightening pollution controls
despite fierce opposition from American automakers.

"America is still in denial about the energy problem and few politicians are prepared to accept painful solutions. We have a peculiar situation in which companies like General Electric and BP are more progressive on energy policy than the administration and Congress."

—James Cooper, Democratic congressman from Tennessee, July 26, 2005

Above: A demonstration of how reliant the modern world is on electricity. This is a mass exodus of people leaving Manhattan over the Brooklyn Bridge when the city suffered a massive power cut on August 14, 2003. Sweltering New York was thrown into chaos as thousands of commuters were stranded and many more trapped in the subway system, bringing back fearful memories of the September 11, 2001, attacks.

"When you start messing around with
these natural systems, you can end up
in situations where it is unstoppable.
There are no brakes you can apply.
This is a big deal because you cannot
put permafrost back once it's gone."

—David Viner, a senior scientist at the Climate Research Unit
at the University of East Anglia, on the melting of the
Siberian tundra, 2006

First of all, there are lots of events that can make the climate warm faster, and in a way that puts reversing the process outside our control. There are also events that can act to cool the climate, again on a scale that dwarfs man's attempts to prevent them. In the jargon they are called positive and negative feedbacks. An example of a positive feedback is melting ice and reduced areas of snow cover. Ice and snow, because they are white, reflect most light back into space. Bare rock, particularly dark rock, and seawater absorb much more of the sun's rays. You only have to sit on a stone wall after sunset and feel the heat on your backside to understand the process. Perhaps a better example is the contrast between black and white cars. A white car with the sun on it remains relatively cool compared with the heat absorbed by a black car.

Over the last 30 years, vast areas of ice in the Arctic and Antarctic have been disappearing. Rock in Greenland and the Antarctic that has been covered by permanent ice for 10,000 years is now exposed to sunlight in the summer and is warming up. The same is true for even larger areas of the ocean around the North Pole.

A second and related problem is the melting of the permafrost. Vast quantities of carbon are stored in the frozen soil, along with large amounts of methane from once rotting vegetation. As it warms, the stored carbon and methane is released into the air, increasing the warming effect. The destruction of forests also leads to ever larger amounts of stored carbon being released into the atmosphere as carbon dioxide. While most of this is man-made destruction, forests are also being damaged by climate change and releasing carbon dioxide as a result. The destruction of forests is estimated to add as much as one-fifth of the carbon dioxide currently being added to the atmosphere by

Opposite: The effects of thawing frozen soil have a profound effect on the locality and the climate. In this picture on the left, taken in June 2002, the Siberian tundra was thawing out. It can be seen in the standing water lying in pools across the landscape. The tundra's permafrost, which is a frozen layer of soil many feet deep, is turning into bogs, ponds, and wetlands. In this image vegetation is dark green and pooled water is dark blue or almost black, and it can be seen in the upper-left quadrant of the image in the Kolymskaya wetlands. Rust-colored areas show bare, exposed soil and, in the upper right, is where spring has yet to arrive in high-altitude terrain. All around the Arctic it is the same.

On this page: As the permafrost thaws, the ground begins to heave as it melts. At the top the effects of melting permafrost can be seen in Alaska in what are called drunken forests. At the bottom graveyard crosses are tilted at odd angles.

burning fossil fuels. Less tree cover also adds to the problems of heat and drought, since trees retain moisture and temper the climate. These positive feedbacks are already adding to the problem of climate change and accelerate the process. Examples of negative feedbacks, which are also difficult to measure, include what is known as the carbon dioxide fertilizer effect—industrial and vehicle pollution and cloud cover.

Plants and trees need carbon dioxide to grow through photosynthesis, so extra amounts of gas in the air, in theory, have a fertilizer effect, making most things grow faster and thus fixing more carbon. In laboratory conditions this can be measured precisely—for example, by growing plants and trees in large greenhouses with increased carbon dioxide in the air. It works; they do grow faster. However, in the real atmosphere, where all the other variables such as moisture and temperature are not controlled, it is a far more difficult effect to measure.

Pollution—sulphur dioxide gas from coal-fired power stations, for example—reacts in the atmosphere to form tiny particles called aerosols, which reflect sunlight back into the atmosphere before it reaches the ground. The effect of aerosols, albeit from a different cause, have been measured in the spectacular aftermath of June 12, 1991, when Mount Pinatubo in the Philippines erupted, injecting 20 million tons of sulphur dioxide into the atmosphere, along with huge quantities of dust. The clouds of material reached 12 miles (19 km) above the Earth. Rather than falling back to the ground, the plumes of the eruption were pushed so high that the material was caught up by the powerful horizontal winds that circulate in the stratosphere, which begins about 7 miles (11 km) above the

Earth. Instead of just having a local effect like smoke from a power station, factory, or city, the dust—particles smaller than one-hundredth of a millimeter—spread around the world in a giant ring, causing, among other things, spectacular sunsets. The amount of radiation reaching the lower atmosphere fell by 2% as a result and the global average temperatures fell by 0.25°C for two years.

By the end of that period, the particles had fallen back to Earth, but for climate scientists it was a gigantic real-life experiment. Another remarkable example occurred after the September 11, 2001, attacks in New York. For three days all air flights in the United States were grounded and daytime temperatures increased, while night temperatures remained the same. Researchers concluded that a lack of aircraft exhaust, an otherwise permanent feature of American skies, had made the difference. The tiny ice crystals in the exhaust normally reflects back the sunlight, an effect previously unrecorded. It has been called global dimming, but the effect of aircraft exhaust and other forms of pollution on clouds is extremely complex. Water vapor is a potent global warming gas, but in the form of clouds it can act both to cool and to warm the climate.

Aerosols from pollution promote extra cloud cover. As anyone who has traveled in an aircraft knows, the dazzling white of the tops of clouds reflects sunlight back into space. Many aerosols have the effect of making clouds appear brighter from space. This has an additional cooling effect by reflecting back the sunlight. Research published in December 2005, jointly by the UK's Meteorological Office and the U.S. government's National Oceanic and Atmospheric Administration, looked at the problem again and concluded that the cooling

Opposite: Aircraft are the fastest-growing contributors to global warming as air travel continues to increase. As well as greenhouse gases, water vapor causes these contrails, seen here over Deal in Kent, England, which help to form high clouds. Scientists are still working on exactly what effect this has on global warming. But the effect of aircraft pollution has been estimated to be about 2.7 times more than that of their carbon dioxide emissions alone because of where they are emitted into the atmosphere. Despite many calls for aircraft fuel to be taxed, the United States has so far vetoed all proposals to do so.

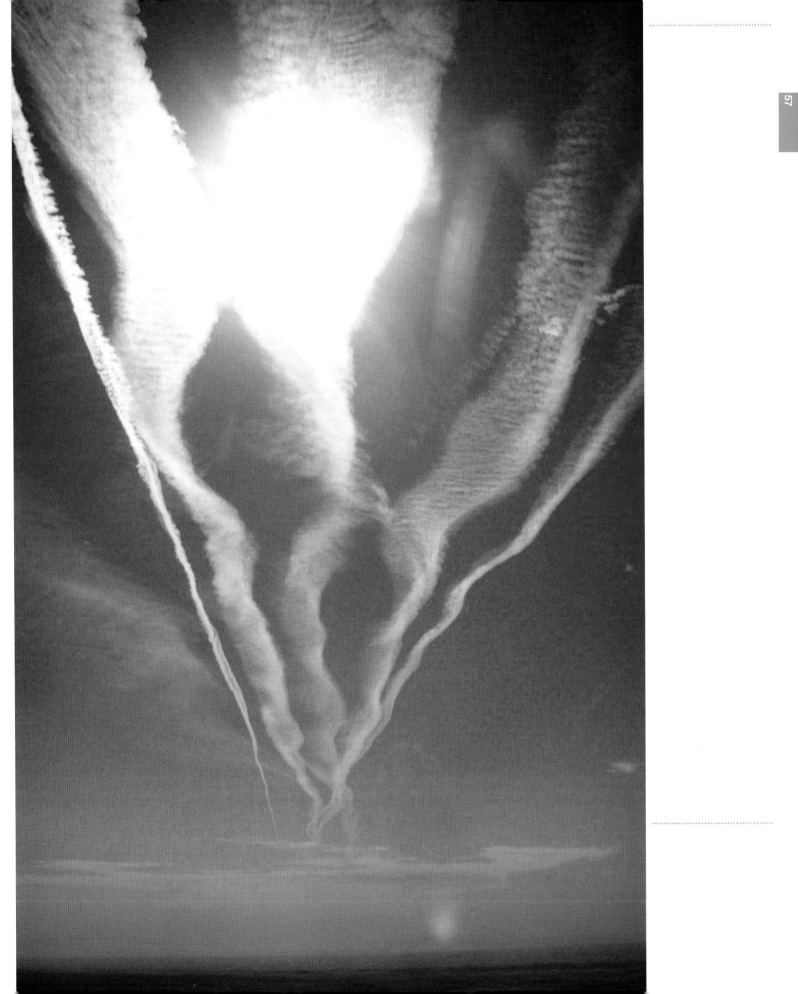

"The weight of evidence for climate change is very strong indeed, and it has gotten stronger over the years since I began chairing the meetings of climate scientists in 1988. The rate of warming is now greater than it has been for 10,000 years; that means the rate of climate change is greater than it has been for 10,000 years."

—Sir John Houghton, the first chairman of the U.N.'s Intergovernmental Panel on Climate Change, 2006

effect of aerosols from factories, forest fires, and dust particles swept up by desert storms, which had been taken into account and calculated, had been badly underestimated.

Clouds also have another effect, acting like a blanket at night to keep the heat trapped at the Earth's surface. It is no accident that in temperate regions clear nights in winter are frosty, and cloudy ones keep the ground temperature well above freezing. Depending on the height of the clouds and the time of day or night, the reflective or blanketing effect of clouds obviously varies greatly. Scientists are still studying this vital area and are trying to discover how different types of clouds at different heights affect temperature and whether changes in cloud cover are keeping the Earth warmer or making it colder.

The latest information is that aviation is a far greater cause of climate change than was previously believed. Not only is it the world's fastest-growing source of carbon dioxide emissions, aircraft also produce water vapor. This condenses to form ice crystals in the upper part of the lower atmosphere, known as the troposphere. These ice crystals, popularly known as vapor trails, trap the Earth's heat. Taken together, the carbon dioxide, local ozone formation from other aircraft emissions, and the condensation trails of the aircraft have 2.7 times the effect of carbon dioxide alone.

These frontier areas of science and the uncertainties that arise because of changing cloud cover mean there is still plenty of room for doubt about how quickly temperatures will rise and by how far. Nearly all of the new evidence points to the fact that the rises will be greater than previously thought. One other vital point involves changes in the measurements of global average temperatures. What has become apparent is that the planet does not warm evenly. The places predicted to get warmer more quickly are the Arctic and the Antarctic, where vast quantities of ice dominate the landscape, and in the Northern Hemisphere, where permafrost prevails. In fact, permafrost covers 27% of the Earth's land surface. These are exactly the areas where the most dangerous positive feedbacks occur and the melting is already occurring.

This is what worries scientists about the average of 2°C across the whole globe. Localized warming in Greenland will not be the average 2°C but is expected to be above 2.7°C, a significant rise because it is above the point that triggers the melting of this vast ice cube. In melting, the 1 million square miles of 2-mile-high Greenland ice cap will raise sea levels by 22 ft. (7 m)—not all at once, of course, but it is unstoppable once the tipping point is reached. Even though scientists believe it could take between 1,000 and 3,000 years before all the ice disappears, the sea-level rise will begin to have impacts much sooner.

In October 2006 the latest measurements of Greenland showed that its glaciers were unloading three times as much ice into the sea as in 1996—a total of 60 cu. miles (250 cu km) a year, enough to cover the southeast of England in 16 ft. (5 m) of water. This meant that previous estimates of Greenland's contribution to sea-level rise had to be revised. The rise due to Greenland's ice losses was thought to be tiny, but current estimates have risen to half a millimeter each year. This may not sound like much, but to scientists it is very ominous. Liz Morris, Arctic science adviser at the Scott Polar Research Institute at Cambridge, said: "It appears Greenland is just on the turning point."

Opposite: Baffin Bay, located between Greenland and Baffin Island, is the site of noticeable climate change. Ever-larger quantities of low-salinity meltwaters from the disappearing ice moving south into the warmer seas are affecting ocean currents. The flow rate of the Jakobshavn Glacier on the western slope of Greenland has increased significantly and is now moving at about 6 miles (10 km) per year, calving a large number of icebergs into Baffin Bay and adding to sea-level rise.

Creating a Scorched Earth

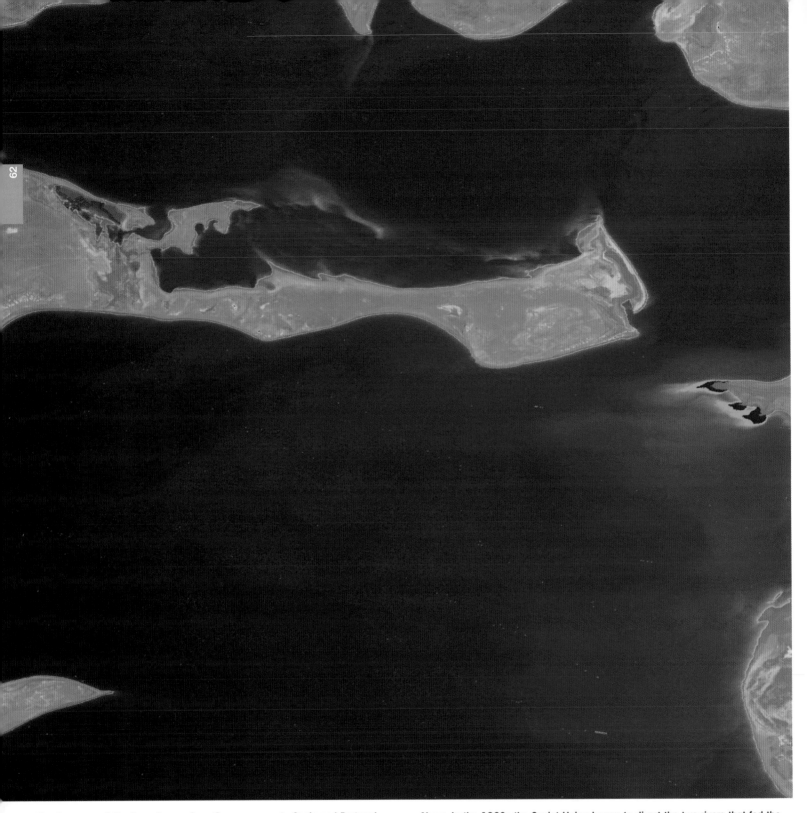

Previous spread: Southern Europe from Greece across to Spain and Portugal has been increasingly plagued by forest fires in the last decade as summers become hotter and drier. In August 2005 firefighters size up the task ahead as a wildfire burns a forest near the village of Bouleternère, near Perpignan, southern France.

Above: In the 1960s the Soviet Union began to divert the two rivers that fed the Aral Sea to provide irrigation so that cheap cotton could be grown in what was barren landscape. The sea, once the fourth-largest lake on Earth, began to shrink. This picture was taken in 1973. By 1989 the northern and southern parts of the sea had already become virtually separated. The drying out of the sea's southern part exposed the salty seabed. Dust storms increased, spreading the salty soil into the agricultural lands. As the agricultural land became contaminated by the salt, the farmers tried to combat it by flushing the soil with huge volumes of water. The water that made its way back to the sea was increasingly saline and polluted by pesticides and fertilizer.

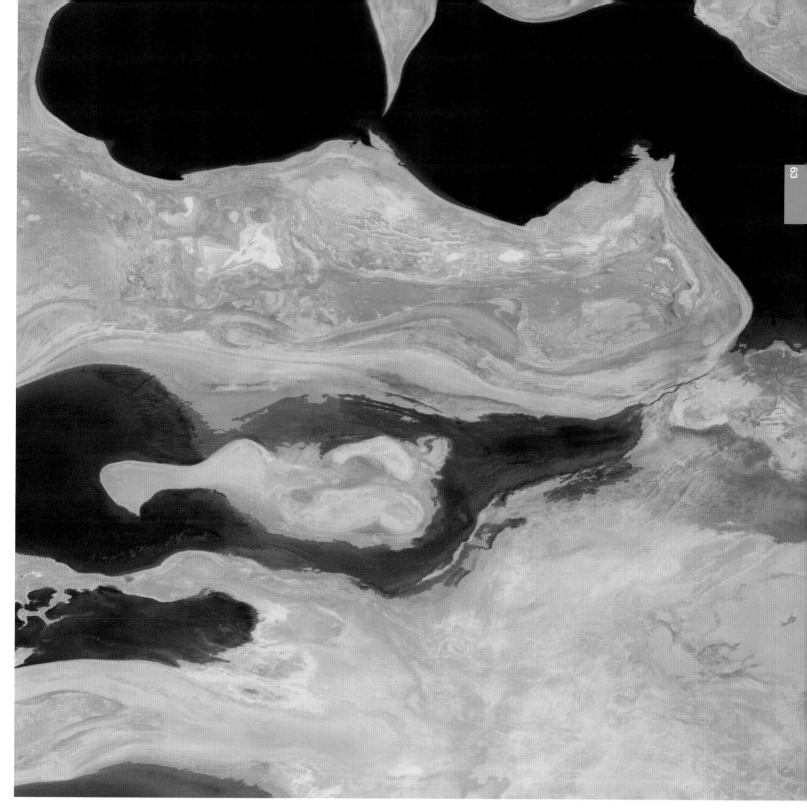

Above: By the time this picture was taken in 2000, the sea was disappearing. The larger southern part, now drying out and separated into a western and eastern half, faces the same bleak future. Complete disappearance could happen in as few as 15 years. The southern Aral Sea has been deemed beyond salvaging, and a restoration effort by the Kazakh government, funded by the World Bank, instead focused on the much smaller, but less polluted and saline, northern sea. A permanent wall was built between the two portions of the sea, sealing the southern half's fate. The northern small Aral Sea has been allowed to refill from the restored inflow of the Syrdar'ya River. It has refilled enough to support a fishery again, and the sea is gradually extending to its former shores. The plan has been so successful that there are plans to build the wall across the sea even higher to contain even more water. The restored northern small Aral Sea will also help stabilize the continental climate and increase rainfall, smooth out winter-summer temperature extremes, and suppress dust storms.

Man-made global warming has a short history and a long future. The natural balance of gases in the atmosphere began to change with the Industrial Revolution, but in the last 50 years, as the whole world has begun to demand a higher standard of living, greenhouse gases in the air have been increasing at an ever faster pace.

It is already clear that the effects of the extra pollutants mankind has released over the last 200 years will be felt by future generations for at least 1,000 years, probably 10,000. As long as fossil fuels are burned at the current pace, never mind the faster speed that most predict, the processes already begun will continue to accelerate and change the planet forever. The comprehension of what man is doing to the planet is relatively new. The French mathematician and scientist Jean Baptiste Fourier was the first to realize that the temperature of the Earth was controlled by the gases in the atmosphere. In 1827 he observed that certain gases trapped heat. Unusual for a scientist, he was able to describe a complex theory so simply that everyone could understand it. He said the atmosphere was like the glass in a greenhouse. It let the sun's rays in and therefore the warmth, but certain gases, notably carbon dioxide, provided a barrier that prevented the heat from escaping again. He even coined the phrase "greenhouse effect."

Not that the greenhouse effect is necessarily a problem—in fact, it is essential for life on Earth. Without this mix of gases that control the heat at the Earth's surface, conditions for mankind would be unpleasant if not impossible. The average temperature has been around 15°C for the last 10,000 years, regulated by the heat-trapping natural gases in the atmosphere, mainly water vapor. Without them the temperature would be well below freezing, about -15°C, soon snuffing out most life.

Fourier's theory was developed by a British scientist, John Tyndall, in 1860. He measured the absorption of heat by carbon dioxide and water vapor and suggested, correctly, that a reduction in the amount of carbon dioxide in the atmosphere might have led to a lessening of the

greenhouse effect and the ice ages. Another 36 years passed before a Swedish chemist, Svante Arrhenius, considered the opposite possibility. What would be the result of significant increases in carbon dioxide in the atmosphere? He believed doubling the concentration of carbon dioxide would increase the temperature across the globe by 5°C to 6°C. Although modern scientists would say it was for the wrong reasons, it is remarkable that despite all of the progress in science in more than a century, and the immense complexity of the climate system, these calculations are very close to what scientists now estimate will happen—within the lifetime of some people now being born.

Although carbon dioxide amounts were already rising, it was not until 1938 that G. S. Callender, a British meteorologist, tried to raise the alarm. He attempted to persuade the august but skeptical Royal Society in London that global warming was already taking place. He had gathered information from 200 weather stations across the world and demonstrated that in the previous 50 years temperatures had increased. He was not taken seriously.

We know now that Callender was very much on the right track, though probably only a small part of the warming in his day was man-made. It was not until World War II had come and gone and science was having a special year—the International Geophysical Year of 1957—that the story moves on. The world's scientists were cooperating to study the Earth's natural systems, bringing America into the story. Two scientists from the Scripps Institute of Oceanography in California warned that the human race was carrying out a huge experiment with the atmosphere and therefore the future of the entire planet. Roger Reville

Opposite, top: A Kazakh villager carries a bucket of water from a well on the former bed of the Aral Sea, outside the village of Karateren, southwestern Kazakhstan in April 2005. Dried salt containing pesticides from cotton crops blows across the desert plains, threatening the health of local people. The sea has retreated 100 miles (150 km). One of the rivers that used to feed the Aral, which runs through Uzbekistan, still has all the water removed for irrigation before it can replenish the sea. The other, the Syrdar'ya, is allowed to flow to the Aral by neighboring Kazakhstan. This will eventually restore this area to a sea.

Opposite, bottom: Children run past ruined ships abandoned in sand that once formed the bed of the Aral Sea near the village of Zhalanash, in southwestern Kazakhstan. One of the rivers that fed the sea, the Amudar'ya, once known as the Oxus, was as large as the Nile, but it no longer reaches the sea, causing what has frequently been called the single greatest ecological disaster caused willfully by man. The once healthy people who lived on a diet of fish from the lake and had fertile fields and gardens have serious health problems, and there is a high instance of birth defects.

Trends in Carbon Dioxide Concentrations

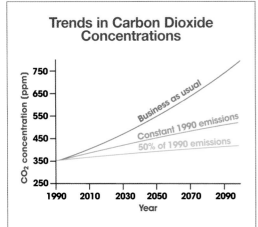

There are three possible trends in carbon dioxide concentrations in the atmosphere, depending on how the human race combats climate change. According to climate scientists, "business as usual" will lead to catastrophe. Their recommendation is a 60% cut from 1990 emissions.

SOURCE: IPCC/Hadley Centre for Climate Prediction and Research.

and Hans Suess thought that the buildup of carbon dioxide in the atmosphere could be dangerous.

The warnings of these two men were at last taken seriously—in fact, ever since the United States has been in the forefront of investigations into the potential effects of climate change and actual measurements of what is happening. As a result of these fears, routine measurements of carbon dioxide in the atmosphere were initiated at one of the most remote places on the planet, the observatory on Mauna Loa, in Hawaii. At 11,000 ft. (3,300 m) above the sea and far away from industrial centers, the air is constantly measured to give scientists around the world a record of the composition of the atmosphere. While it always shows a summer and winter fluctuation in carbon dioxide levels, which ties in with the seasons and the absorption of the gas by plants, the trend has been constantly upward year to year for half a century. The level has been increasing faster and faster with each decade. The rising measurements reflect the ever increasing burning of fossil fuels and the greater destruction of forests.

In view of more recent history, it is richly ironic that it was the U.S. scientific community and its politicians that first demanded action to prevent the threat of climate change. This thirst for scientific knowledge and the politicians' acceptance of the need to respond to the grim warnings that began in the 1950s did not change until the late 1980s and 1990s. It was at this point that U.S. politicians began to shrink from even appearing to threaten the comfortable lifestyles of their electorate.

The tackling of climate change and its potential inconvenience for voters was not an issue before then. This was partly because in the 1960s and '70s there had been a relatively cold

period in our recent climate, and the prospects of rapid global warming seemed remote. There had even been talk in the 1970s of the return of the Ice Age, an episode still frequently quoted by global warming skeptics in an attempt to discredit current climate science.

While nothing adverse appeared to be happening to the weather, despite the fact that measurements showed carbon dioxide continuing to build up in the atmosphere, scientists continued to be concerned. The World Meteorological Organization, now better known by its initials WMO, sponsored a conference on climate fluctuations at the University of East Anglia, Norwich, England, in 1975. This is still a center of excellence in climate science. The U.S. National Academy of Sciences was also not letting the matter rest and, in the same year, produced a disturbing report called Understanding Climate Change: A Program for Action. The report was concerned about the probable warming effect on the Earth because of the emissions from heavy industry. Two years later, in a second report, the academy went further and warned that the implications of projected climate change "warrant prompt action."

Partly as a result, the first World Climate Conference took place in Geneva in 1979. Scientists noted the increased carbon dioxide in the atmosphere and attributed it mainly to increased industrialization, raising for the first time the loss of forests as a cause. The cutting down and burning of forests released stored carbon in the wood back into the atmosphere, the scientists said. A year later a follow-up conference estimated that a doubling of carbon dioxide levels in the atmosphere would cause global warming of between 1.5°C and 4.5°C, an underestimate by current thinking.

Opposite: The graceful domes of an orthodox church in St. Petersburg seen against the smokestacks and sunset of a winter sky. This picture was taken to illustrate the challenge the world faces in tackling global warming. Russia, after months of prevarication, caused when its scientists claimed that global warming might be good for such a cold country, finally ratified the Kyoto Protocol in November 2004, enabling it to come into legal force on February 16, 2005.

Additional meetings in 1985 and 1987 reinforced these findings and suggested an international treaty to cut back on the rate of greenhouse gases in order to reduce the dangers of excess warming. The following year the WMO and the U.N. Environment Program set up the Intergovernmental Panel on Climate Change, the IPCC, to assess scientific information and formulate response strategies. In June 1988 another significant event occurred: James Hansen, from NASA, testified live on television that he was 99% certain that the warming in the 1980s was not a chance event but linked to climate change. He said: "It is time to stop waffling so much and say the evidence is pretty strong that the greenhouse effect is here." He was speaking at the same time that the United States was suffering a catastrophic drought in the Midwest, which he appeared to be blaming on global warming. At the time, his comments created a sensation and were reported around the world.

In the same month in Toronto, 48 countries took part in a conference entitled The Changing Atmosphere: Implications for Global Security. The conference called for a 20% reduction in global carbon dioxide emissions from 1988 levels by 2005, with the eventual aim of a 50% reduction. The conference even talked of a carbon tax to try to reach these aims. These conference targets to reduce greenhouse gases seem in tune with current scientific thinking of what needs to be done. What was wildly optimistic, in light of recent history, is the time frame by which those present thought these targets could be achieved. It was clear, though, even in 1988 that the threat posed by environmental damage to the planet had reached the political mainstream.

Sir Crispin Tickell, a scientist and senior diplomat, has been a tireless campaigner warning about the dangers of climate change for more than 20 years. In the 1980s, in the middle of a distinguished career in which he was British ambassador to the United Nations among many other jobs, he was close to the center of power. He had studied climate change at Harvard University and is credited with convincing former British prime minister Margaret Thatcher of the seriousness of environmental issues, particularly global warming, while sitting next to her on an airplane. He also helped write a key speech for her on the issue. Margaret Thatcher, then at her highest profile on the world stage, spoke to the Royal Society in September 1988 about the issues of acid rain, damage to the ozone layer, and the threat of global warming. It was a message she later repeated to the United Nations in New York.

So there was an apparent international acceptance of the problem and, as the conference in Toronto illustrated, already a comprehensive understanding of what action was necessary to deal with the threat. But nothing substantial actually happened. From that moment, it seems, the issue of climate change moved from the purely scientific to the political. The IPCC has become the biggest collaborative scientific effort in history, but it is also the center of an enormous political circus. Every country in the world is entitled to contribute its expertise. Some, particularly oil producers and coal users and exporters who regard the possible curtailment of the use of fossil fuel as a threat to their economies, pack the proceedings with as many climate skeptics as possible. These are the professional contrarians. They are part of a sophisticated and very successful campaign funded by the fossil-fuel lobby to prevent action on climate change, designed to destroy and discredit the science and prevent political action to reduce

Right: Sometimes mystified motorists in northern Europe notice that their cars have acquired a rust-colored tinge caused by red dust. This picture, taken in April 2003, captures the source of the pollution—minute particles of sand whipped high into the air by powerful desert winds in the Sahara. The dust can discolor the snow on the Scandinavian mountains, reducing its reflective power and allowing it to absorb heat and therefore melt.

"It is crazy for us to play games with our children's future.... We have a heavy obligation because we now know what is happening to our climate and what will be a highly predictable set of outcomes if we continue to pour greenhouse gases into the atmosphere. We also know we have an alternative of even greater prosperity if we apply the new technologies that are now available to us."

—Former president Bill Clinton, speaking at the Montreal Climate talks, December 9, 2005

greenhouse-gas emissions. More details follow in the next chapter, but a contrarian can argue about any point for hours—for example, whether the phrases greenhouse effect, climate change, and global warming should be used in official documents. The greenhouse effect, being a natural phenomenon, should properly have the word "enhanced" in front of it to make clear the role man has in causing global warming. On the other hand, contrarians do not accept that the globe is warming, or at least that man is responsible, so that phrase is out altogether in favor of climate change, which means the temperature could go down as well as up. Journalists scoff at such pedantry, but the contrarians have held up whole conferences for hours with such empty discussions.

It is not that scientists, in spite of these arguments and obstructions, ever stopped looking at the real problems. The creation of the IPCC and its subsequent series of reports on the causes and the effects of climate change have been detailed and comprehensive. A vast body of new science on the climate has been created, nearly all of it reinforcing the original view that the world is facing a crisis.

As it has become clear that a rise in sea level is already inevitable and temperature rises are already built into the system, increasing emphasis has been placed on the need to adapt. It is already accepted that climate change is unstoppable. Estimates of the economic costs of doing nothing about cutting emissions have been increasingly alarming. Perhaps the most important single conclusion of scientists is that man is clearly responsible for the problem. What they are saying loud and clear is that only political action by world leaders can hope to stop climate change from becoming a catastrophic disaster for both the natural world and

the human race. Sadly, some scientists feel their job is done, but more should take the example of Sir Crispin Tickell and do their civic duty and reinforce the message to politicians and the public as often as possible. However, even without that extra push to public opinion, there can be no doubt that the message has been conveyed frequently enough to the world's political leaders for the crisis to be unmistakable.

In fact, it is quite clear that in some senses the message has sunk in, because while it appears that despite these unmistakable warnings that little has been done, it is not because politicians have stopped talking about the problem. There has been an unprecedented series of high-level meetings on the environment at which climate change has usually been at the top of the agenda, and sometimes the only topic. A list of them would run into the hundreds, many of them appearing to have achieved very little. But international diplomacy is like that— incremental rather than dazzling breakthroughs. There have been some important milestones along the way, all providing an international framework so that the world leaders could take action on global warming if they could summon the political will.

General concern about the degrading of the Earth's environment was the impetus behind the 1992 Earth Summit in Rio de Janeiro. A centerpiece for the summit, then the largest the world had seen, was the U.N. Framework Convention on Climate Change. It is a remarkably short and clear document, rarely quoted, but in light of subsequent events, worth quoting. The "objective" of the convention is but a single paragraph. It "is to achieve ...stabilization of greenhouse gas concentrations in the atmosphere at a level that would prevent dangerous anthropogenic (man-made)

Top left: Another "historic moment" in the climate negotiations. The Japanese environment minister, Hiroshi Oki, is seen on two screens as he addresses the plenary session of the U.N. global warming conference in Kyoto on December 11, 1997. This conference was where the protocol—aimed at heading off a potentially catastrophic warming of the Earth and now known by the city's name—was formally adopted after 11 days of tense debate.

Top right: After the United States walked out of the climate talks yet again, this time in December 2005, Bill Clinton, the former president, rose to speak, calling on his successor George W. Bush to come back to the negotiating table to safeguard the future of the planet.

Bottom left: While Bill Clinton uses his persuasive powers inside the conference hall, thousands of people march outside through the streets of Montreal in December 2005 as part of a worldwide day of protest against global warming.

Bottom right: A Greenpeace member protesting at the Bush administration's repudiation of the Kyoto Protocol at the climate talks in Bonn in July 2001. It was the meeting at which the rest of the world decided to proceed with the legally binding agreement and ignore America's attempts to destroy the treaty.

interference with the climate system. Such a level should be achieved within a time frame sufficient to allow ecosystems to adapt naturally to climate change, to ensure that food production is not threatened, and to enable economic development to proceed in a sustainable manner."

The first key principle on the next page says: "The parties should protect the climate system for the benefit of present and future generations of humankind." It goes on to say that the developed countries, those that had already grown rich on polluting the atmosphere, "should take a lead in combating climate change and the adverse effects thereof." Many nations signed up to it immediately. As the key principles go on to explain, the convention puts the onus on those who had signed and ratified the agreement in their parliaments to "adopt policies and measures" to achieve a safe climate.

Countries agreed as a first step that by the year 2000 they would get carbon emissions back to 1990 levels, a promise all the politicians apparently forgot the moment they signed up to it. In any event, they did nothing to make good on their pledges. The lesson learned was that countries were obliged in the treaty to take measures to tackle climate change, but in effect the convention was rendered fruitless because the target was not legally binding. It took more than 10 years to rectify this problem with the adoption of the Kyoto Protocol. The 1992 summit was important for other reasons, both for the agreements it reached and for those it failed. Another success was the Biodiversity Convention, designed to prevent the continued extinction of animals and plants. It was also agreed and signed, although the conspicuous reluctance of the United States to take part because it might cost too much money and

Left: The clear cutting of the Kenogami boreal forest in Northern Ontario, Canada, east of Thunder Bay between Terrace Bay and Geraldton, is mostly for the production of facial tissues. This has enraged environmental groups. The forest is one of the last strongholds of the endangered woodland caribou, gray wolves, black bears, bald eagles, red foxes, American white pelican, owls and songbirds that return each spring. Nearly 100 miles (160 km) of new logging roads are being built through the forest each year as more and more of the 4.9 million acres (1.98 million hectares) are cleared for wood pulp.

affect American interests was an ominous indication of what was to come. A forest convention, billed by Europe and the United States as a major object of the conference, never materialized. The developing world refused to accept any interference with the sovereign rights to control their own natural resources.

Fifteen years later a forest convention is no closer to fruition. One of the surprises of the conference was the move by developing countries to create a convention to combat the spread of deserts. Although it was barely of interest to the developed world, which apart from the United States and Australia does not have major deserts, the convention moved forward quickly. The subsequent program of international cooperation and measures to push back desert frontiers has been the least reported but one of the most successful outcomes of the Rio summit. There was also an extremely long and mostly unread document called Agenda 21, which was a blueprint for the 21st century, designed to get the planet on course for a sustainable future. Many local, rural, and city governments have subsequently taken its provisions seriously and made a huge difference to the quality of local life. A far greater number have ignored it.

The whole summit ended in a mood of optimism. At last there was a feeling that there would be international action to preserve the environment. Even though there were tensions between the developing countries and the rich nations, everyone was heading in the right direction. There are a number of important themes that emerged which are as influential today. The developing countries' point, still argued strongly, is that the rich countries grew rich by wrecking the environment, and since they had

Left: The tribal lands of Yanomami Indians of the Roraima region of the Amazon are supposed to be protected areas, but logging continues anyway. The loggers are followed by farmers who clear the remaining forest. Although the Amazon is still an area of tropical forest larger than Europe, it is being eaten away at an ever growing rate. The Amazon is known as the lungs of the world because of the amount of carbon dioxide the forest absorbs. This process, which is helping to keep the climate stable, is being lost as the trees are destroyed.

"The last 50 years stick out like a sore thumb.... The temperature's gone up and up and up. It bears the imprint of human activity."

—Dr. Michael Oppenheimer,
professor at Princeton University,
speaking on *Oprah,* November 2006

caused the problem they should bear the brunt of fixing it. Meanwhile the developing countries' priority was, and is, to continue developing. This was the principal reason for the foundering of the forest convention. As has already been said, developing countries regarded their forests as their own natural resource and if they wanted to cut them down, either for cash or for development, it was their sovereign right to do so. It could be argued that some rich countries, Canada and Russia for example are continuing to do so. Canada has only 16% of its southern boreal forests intact and timber companies continue to press north into primary forest. Less than 7% of Canada's land area is protected from development. Thinking on how to protect forests and the importance of leaving them standing is beginning to change but in 1992 developing countries were pointing out that European countries had long ago cut down their own forests for agriculture and for building materials for boats and houses, so were in no position to criticize poorer countries for doing the same. One significant point about all these agreements and negotiations, successful or otherwise, was how they were interrelated. There was the overall theme of how to develop while at the same time avoiding the destruction of the environment. But just as important, as we see in the science of climate change, all these environmental matters are related. How can climate change be considered without also looking at the loss of forests, the spread of deserts and the disappearance of so many plants and animals?

The optimism of the Earth Summit proved to be a false dawn, partly because the world was in the middle of a mini recession, where jobs were being lost and Western economies were stagnating. The environment took second place to the need to stimulate growth. There is also a time lag built into international treaties, which involves waiting for enough governments to pass legislation through

their parliaments to bring any convention to fruition. In the case of the Climate Change Convention, this was remarkably swift compared with other attempts at international regulation, which have sometimes languished for more than a decade before reaching a critical mass with enough participating nations prepared to make parliamentary time to bring them into force. In the case of the Climate Convention, it took only three years before 50 of the 150 countries that had signed it in Rio ratified it, opening the door for the first Conference of the Parties (or COP 1) in Berlin in 1995. Under the convention rules, once started, these COPs are held every year. They now stretch over two weeks from the end of November into the first week of December, usually with a full attendance list of politicians, indicating their importance. Midway through the year in June, a lower-level meeting of civil servants is held to assess progress from the last meeting and prepare the ground for the next. The landmark meeting in Montreal in December 2005 was COP 11, but more about that later.

The result of the first COP, the Berlin meeting, was a mandate to seek a legally binding agreement within two years to cut greenhouse gas emissions. It was already acknowledged that the voluntary agreement, reached in Rio but ignored, was never going to deliver a safer climate. One of the features of the Berlin meeting, and previous political gatherings about climate change, was the emotional but completely reasonable appeal of the Alliance of Small Island States (AOSIS). This was the first major grouping of countries concerned about climate change. They were the first to realize that the theoretical threat of sea-level rise, and storm surges had for them become a present danger. Many of these states, like the Maldives and the Marshall Islands, are based on vulnerable coral islands. Others, like the Seychelles or the

Opposite: Thunderstorms are common in the tropics, particularly over the forests of the Amazon. A NASA-funded study has shown that tiny airborne particles of pollution caused by forest fires modify developing thunderclouds by increasing the quantity of water droplets and reducing their size at the same time. Researchers believe that rainfall over the forested areas is reduced by 20%, enough to cause considerable changes in the tree cover below. The immense forest is home to thousands of plant and animal species—including the elusive jaguar—many of which are still to be recorded. Continued logging threatens its future and that of many other species. Thousands of plant and tree species that occur nowhere else on the planet will also be lost forever. Many have medicinal properties known to the indigenous population but not yet investigated by modern scientists.

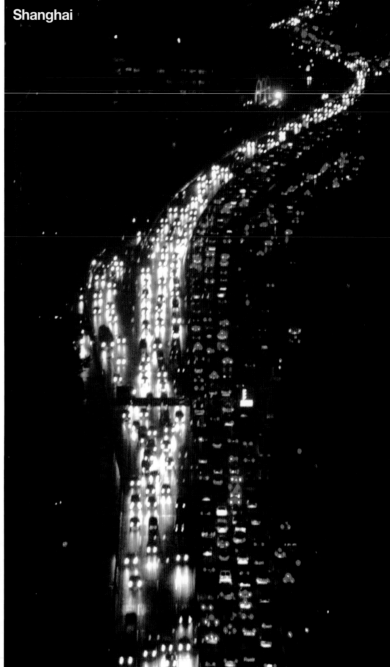

Shanghai

Los Angeles

São Paolo

Dhaka

"Will we stand by while drought and famine, storms and floods overtake our planet? Or will we look back at today and say that this was the moment when we took a stand? That this was the moment when we began to turn things around?"

—Senator Barack Obama

Caribbean islands, have some high land but are already losing beaches on which their main foreign currency earner, tourism, depends. For many of these 36 AOSIS states, increasing storms and weather-related disasters mean, at a minimum, economic instability, and for others, in the worst case, sea-level rise means the end of their existence as nations. Whole populations will have to leave their homelands forever and seek sanctuary elsewhere.

At many of the international conferences where world leaders meet, this impending crisis for these most vulnerable countries has been stridently and often emotionally stated by various prime ministers and presidents of the countries involved. They have been politely listened to and loudly applauded but, in terms of action, largely ignored. As early as 1990, when AOSIS was originally formed, the group had clout in the U.N. system because of the one-nation, one-vote principle. Aided by a handful of young and enthusiastic international lawyers, the group helped to draft the original Climate Change Convention and raised the issue of the inability of these vulnerable low-lying nations to get flood insurance and addressed the crucial question of what would happen to their peoples when sea levels rose. They were an important counterbalance to the obstructions of the fossil-fuel lobby and oil-producing countries like Saudi Arabia.

In Berlin the AOSIS demand at the meeting for a 20% cut in carbon dioxide emissions by industrialized countries by 2005 was specifically included in the mandate. The culprits—the industrialized countries—and the need for the governments of those states responsible to sort out the mess, were again clearly spelled out.

Two years later, as the so-called Berlin Mandate had demanded, a third and, as it turned out, highly dramatic meeting took place in Kyoto. There had been many tortuous preparations but the 10,000 people who gathered in Japan were in for a grueling time. Through all the jargon, horse-trading and lobbying from all sides, the deal was about how much each of the industrialized nations would agree to reduce their carbon emissions. Everyone was prepared to accept that every country had a different set of circumstances. Some were at an early stage of development, some at a late one, and some had special industries that used a lot of fossil fuels. Every country seemed to have a case that it should have a less demanding target than its neighbor.

The most pressure was on the United States, by far the world's biggest polluter, to take responsibility for the mess. The figures speak for themselves. A country with about 4% of the world's population was and still is responsible for 25% of the carbon emissions. With Bill Clinton claiming to be a green president, and his vice president Al Gore the author of a book on the dangers of climate change, it was seen as politically impossible for the United States not to do a deal. In Kyoto, however, hope seemed to fade. After two weeks of tough talking, with the conference due to end in two days and no deal in sight, it was decided to continue night and day in continuous session until an agreement could be reached. Forty-eight hours later the conference was still in session; people were literally dropping asleep in their chairs, and television crews and reporters slept in corridors.

Long after the conference was supposed to have finished, a deal was done. Thirty-six

Opposite: Car ownership is increasing all over the world, and there are few cities anywhere that do not have traffic jams. As this packed freeway in Los Angeles shows, there is never enough space for all cars regardless of how many roads you build. In the booming city of Shanghai in September 2005, the lights on the Yanan elevated road show that China is fast catching up to the developed world in car ownership. Makers from all over the world are trying to break into this number one growth market, but it is not clear whether there is enough room for all the vehicles. In Dhaka, the capital of Bangladesh—one of the poorest countries in the world—where only a very small percentage of the population can afford a car, total gridlock and smog are already part of daily life. Teeming São Paolo in Brazil cannot keep up with its booming traffic, either.

"An increase of two or three degrees wouldn't be so bad for a northern country like Russia. We could spend less on fur coats and the grain harvest would go up"

—Vladimir Putin, Russian president, October 2003

industrialized countries, including crucially the EU, Japan, and the United States, had agreed to legally binding targets to reduce greenhouse gases from 1990 levels by 2008–12. The rather odd four-year period was the subject of much negotiation because some countries thought they could do it at the beginning and others not until the end. No targets were set for developing countries. Industrialized countries would reduce their collective emissions of greenhouse gases by 5.2% compared to the year 1990. That may not seem like much, but the U.N. Environment Program noted that it was a 29% cut on what would have been expected by 2010 without the protocol.

The bottom line for individual countries reflected the horse trading. The United States had agreed to a cut of 7%, Japan 6% and the European Union 8%. In fact, the EU figure hid a complex deal. Some countries in the 15 thought they could do considerably better, notably Denmark, which had a large renewables program, and a unified Germany, which had already closed down a lot of energy-intensive industries in the former East Germany. Both were prepared to take a 21% cut. The United Kingdom, which had closed a lot of coal mines and was transferring energy production to more efficient gas-fired power stations, promised 12.5% cuts.

This provided the less developed countries of southern Europe some increases, with Greece allowed 25% and Portugal 27%. Overall, this allowed an impressive 8% cut for the EU without reducing growth. This is proving to be a tough legally binding target to meet, with considerable doubt that it will be achieved. Most of the former Soviet bloc countries, apart from Russia, followed the EU with an 8% cut, partly through solidarity but mainly because they, too, had lost inefficient industries with the collapse of the communist

regimes. Switzerland, which had no reason apart from desperation about the effect of climate change on the Alps, opted for a difficult 8% cut because it saw the size of the threat to its tourist industry.

Russia, with its vast forests and collapsed industries, was allowed to get away with holding emissions at 1990 levels. This meant that Russia would easily meet its targets, and this was interpreted by the environmental movement as a fiddle in the making. Under the terms of the Kyoto agreement, it was possible for countries that exceeded their targets to sell the tons of carbon saved to those that had failed. This system is designed to reward financially those that have done well and allow those that have failed to buy their way out of a problem. It was felt that Russia was getting a free ride, and it would be a way for the wealthy United States to purchase "hot air" to avoid taking action at home.

Russia is still well placed to exceed its target and cash in on the surplus carbon, but its main potential customer, the United States, has vanished, having repudiated Kyoto. Judging by the failure of many countries to take enough action to curb emissions and meet their targets, Russia will still not be short of customers by 2012. Some countries were allowed rises, as much as 10% in the case of Iceland. Australia, with its large fossil-fuel industries, insisted on plus 8%, a curious anomaly as it turns out. Australia, having repudiated Kyoto, now seems likely to reach its target through industrial changes that would have happened anyway. It shows that predicting the future is an inexact science.

At the signing of the Kyoto deal, the world heaved a sigh of relief—at last something was being done. In fact, since Kyoto, negotiations

Opposite: A steel worker in Ukraine's eastern city of Donetsk, one of the driving forces behind the former Soviet state's reviving economy. With its iron ore and coal reserves, it is well placed to take advantage of high prices caused by steel shortages, but this is at a heavy cost to the atmosphere in terms of increased carbon emissions.

Above: The prospect of reviving the Russian economy by reopening some of the older, less-efficient heavy industrial plants of the Soviet era were said to contribute to President Vladimir Putin's apparent reluctance to sign up for the Kyoto Protocol in 2004. He was concerned that the Russian government would not be able to keep its pledge to keep emissions below the 1990 levels and grow its economy at the same time. In the end, trade inducements from the European Union overcame other factors.

on climate have been painful and essentially depressing, although not without excitement. The high spot in drama was when the newly elected president, George W. Bush, repudiated the Kyoto agreement exactly three years after it had been agreed. This was partly for ideological reasons but also because the U.S. economy had grown so much since 1990 that America could not hope to reach its targets. As President Clinton pointed out in his hard-hitting speech at Montreal during a crucial moment in the 2005 climate talks, he had in his last years as president tried and failed to bring in legislation that would reduce America's emissions.

It had become clear to the Bush administration that the United States's production of greenhouse gases had increased so much during the prosperous years of the late 1990s that it was already impossible to reach the 7% reduction target, even if there was an ounce of political will to do so. But even before the election, it was clear that the oil man, George W. Bush, was the preferred choice of the fossil fuel lobby, which heavily financed his campaign. The speed with which he repudiated the protocol on taking office was the only surprise.

But just when many thought all was lost, and perhaps uniquely in international relations, the rest of the world decided to ignore the world's richest and most powerful nation and continue with Kyoto. Six months later, at what would have otherwise been a routine meeting of civil servant advisers the politicians turned up in force in Bonn, the headquarters of the climate change convention secretariat. There was another exciting moment in the history of the convention, when the remaining industrialized countries that had targets refused to let the agreement die. In effect, at least as far as climate change was concerned, they made the United States an international pariah.

In the following years, the Bush administration has done its best to undermine Kyoto. The convention's main champions have been the European Union and Japan. Apart from other motives, Japan has an emotional attachment to the treaty. Since it was signed in a Japanese city, in addition to any other reason, it is a matter of honor and pride for the nation to make it work.

In the six years since the high point in Bonn when Kyoto was rescued, the political process has stalled. This was partly an inevitable consequence of the sluggish legal process. By the end of 2003, 120 countries had ratified it, which under the terms of the protocol was well in excess of the 55 needed, but there was another provision more difficult to fulfill. This was regarding the emission targets that the protocol had set. Among those ratifying the convention, there had to be enough of the industrialized countries with targets to reduce emissions to represent 55% of the total greenhouse gases released by the industrialized world. Now that the United States, with 25% of the emissions, was out of it, this meant that all the other big polluters had to ratify to bring the protocol into force.

At the end of 2003, the total emissions of the ratifiers was only 44%, which included the EU, Eastern Europe, and Japan. Russia's 17% was needed to reach the necessary total, and the Kremlin was playing hard to get. President Vladimir Putin had more than once said Russia would ratify but some influential Russians from the Academy of Sciences resisted the idea. A bit like some of the climate skeptics in the United States, they poured doubts on the science, but behind the scenes the motive appeared to be a throwback to the cold war. There was some evidence that climate change would be an advantage to Russia, allowing wheat and other

Right: Climate change was at the top of the agenda for the 2007 summit of the G8 leaders in Heiligendamm, Germany. The leaders of the industrialized countries seemed happy enough as they head off for the traditional photo shoot but there were tense exchanges as President George Bush refused to accept a 50% greenhouse-gas-reduction target demanded by the United Kingdom and Germany. Some heralded President Bush's statement that substantial cuts in greenhouse gases were needed as a major advance, but in reality no concrete actions were agreed upon. Left to right are Canadian Prime Minister Stephen Harper, European Commission President Jose Manual Barroso, British Prime Minister Tony Blair, Russian President Vladimir Putin, French President Nicolas Sarkozy, Japanese Prime Minister Shinzo Abe, German Chancellor Angela Merkel, Italian Prime Minister Romano Prodi, and U.S. President Bush.

crops to grow farther north than now. Changes in rainfall patterns would also benefit some areas, while climate change was clearly more detrimental to the old enemy, the United States. Disadvantages to Russia, including the massive melting of the permafrost and the loss of some forests, were discounted against these advantages. The argument in Moscow was that Russia should not be in too much of a hurry to slow down climate change.

In the end, after using Kyoto as a bargaining chip to extract some backing for trade deals from the European Union, President Putin kept his promise, and the Russian parliament ratified the agreement. Kyoto came into force on February 16, 2005, to much international rejoicing. (The word historic cropped up a lot.) Outside the conference halls the struggle to reach critical mass for ratification had grabbed a lot of attention. There was a lot less interest in the equally important squabbling between the parties over the details of the convention. The obvious way to cut emissions is to burn less fossil fuel, but there were details of various plans to negotiate, including those to introduce clean technologies to other countries in exchange for carbon credits.

One of the most difficult discussions, which went on for years and will continue, is about the rules of how various greenhouse gases could be reduced by growing or replanting forests. There is a debate about how much carbon dioxide is emitted from the ground when it is disturbed to plant new trees, and how much carbon is absorbed by the saplings. In the longer term how much can these totals count against any carbon-emission targets when forests can be burned down, cut down, or subsequently harvested?

The United States and its one ally, Australia, while remaining outside Kyoto, continued to do everything in their power to undermine the political process, including raising difficulties in all these discussions. As time passed, the next phase of what to do about climate change beyond the expiry of the Kyoto Protocol in 2012 began to loom. Efforts were made by Tony Blair, the British prime minister, among others, to woo the White House, because it is clear that without the world's largest polluter in the process, attempts to save the climate will fail.

At the same time, the rapidly industrializing developing world, notably China, India, Brazil, South Africa, and Mexico, were increasing greenhouse-gas emissions at an alarming rate, and were not bound by targets under Kyoto. Their involvement in the process had to be increased so that they can become more active participants in reducing their own emissions as soon as possible.

With Britain's turn as president of the G8 group of industrialized countries coming up in 2005, Tony Blair made a bold move. He announced that he was making the end of poverty in Africa and climate change his twin priorities. As part of this, the leaders of the big five developing nations named above were invited to the G8 meeting at Gleneagles, Scotland, in July 2005, specifically to discuss climate change. They accepted, and progress was made on debt relief for the world's poorest nations, but critics said no concrete progress was made on climate. This was slightly uncharitable since President Bush did make some statements about climate change being a threat. This had long been acknowledged almost everywhere outside the White House, but nonetheless, for the United States it was progress of a sort.

Left: Traffic and pedestrians hustle through Shanghai's Nanjing Street, famous throughout Asia for its shopping. China, and particularly this bustling port city, has embraced consumerism and the Western way of life. With its large port and commercial area only 6.5 ft. (2 m) above sea level, the city is in danger from sea-level rise and is already spending huge sums on building its defenses against tidal surges caused by ever worsening typhoons in the region.

"Unfortunately, some parties, by not abiding by their commitments, have put the credibility of the Kyoto Protocol in question."

—Rafiq Ahmed Khan, high commissioner of Bangladesh, December 2005

Barely a month later, however, when news began to leak out that the Bush administration had done a deal behind Blair's back to share new technology with China, India, Japan, South Korea, and Australia, suspicions about U.S. motives over climate were rekindled. This agreement was seen as another attempt to undermine Kyoto, although all the countries involved have subsequently denied that. Japan's involvement meant that at least the government in Tokyo was sincere in believing it was an extra agreement rather than a replacement. And at Montreal, China's statement that Kyoto was the only credible political agreement to tackle climate change has removed any lingering doubt about its status.

If anything, while progress in reducing emissions has been painfully slow, the political rhetoric and apparent alarm at the imminent threat has increased dramatically. The phrase first used at the beginning of 2005 by Sir David King, the United Kingdom's chief scientific adviser that "climate change is a greater threat to the world than terrorism" at the beginning of 2005 has been repeated many times since by politicians, including Tony Blair, to the point where it is already a cliché. Yet nearly 20 years after the Toronto conference the target of a 20% reduction of carbon dioxide levels of 1988 has not yet been attempted. In fact, total world emissions have gone up at an ever faster rate since then, and according to present trends that seems likely to continue for decades.

It would be unfair to say there has been no progress. Considering the pace of other international agreements, there has been a series of advances in terms of international treaties and agreements. The level of education on the subject at all levels of society has advanced at great speed, not least because the changes are already noticeable and widely reported. It would be fair to say the general public across the world is now aware of how the climate is changing.

In Europe, particularly Britain, where talking about the weather is often the conventional start of any conversation, any unusual variation is attributed to global warming. Hosts of people now monitor the first frog spawn of spring, when birds begin to nest and when leaves sprout and fall, all in an effort to keep track of climate change. Hardly a day goes by without details of new science related to global warming being widely reported. Books are published across the world explaining how to green your lifestyle. Anything from a dozen to a hundred ways to save the planet are recommended in newspapers and magazines.

Yet the political effort to do something about the causes of climate change is at best slow and ponderous. At worst it is fiddling while the Earth burns, and it is an awful abdication of the responsibility of leadership.

At the end of 2005, eyes then turned to the COP 11 meeting in Montreal. This is where the world's experts in the politics of climate change met to explore the future. It was also the first meeting of the parties to the Kyoto Protocol, the so-called MOP 1. What happened there, at least partly, determines the future of the world as we know it. This was hard to discern from the newspaper and television coverage. The meeting ended with some expressing delight and hope. Others were not so sure. Montreal, and where we go from here, is examined in upcoming chapters.

Left: A motorcyclist and his passenger turn back due to intense heat as they pass through haze near burned peat land in Indonesia's Riau province. These fires in August 2005 were typical of those that beset the region as drought and development encroach on forests and peat bogs. In some years the dense smog has choked areas over hundreds of square miles.

The Kyoto Protocol

The Kyoto Protocol was agreed upon in 1997 and came into force on February 16, 2005. Originally 36 industrial countries signed up and established targets to reduce or control emissions. Two, the United States and Australia, later dropped out. The targets of the 15 member states of the European Union of the time were set jointly so that some EU countries could have higher targets than others. This would allow newly joined southern European states and Ireland, which were regarded as underdeveloped, leeway to increase emissions. To make up the difference, Germany, Denmark, and the United Kingdom all agreed that they could reduce their emissions significantly more than 8% to make up the difference. Most eastern European countries decided to have the same target as the EU, partly because most of them wanted to join the union.

The agreement was to reduce or control emissions of six greenhouse gases. The 1990 level of emissions was used as a base line, and the dates that the target had to be reached was any time between 2008 and 2012.

-8%	The 15 European Union countries, plus Bulgaria, Czech Republic, Estonia, Latvia, Liechtenstein, Lithuania, Romania, Slovakia, Slovenia, and Switzerland.
-7%	The United States agreed but later dropped out.
-6%	Canada, Hungary, Japan, and Poland
-5%	Croatia
0%	New Zealand, Russian Federation and Ukraine
+1%	Norway
+8%	Australia (later dropped out)
+10%	Iceland
	Kazakhstan, which was not an original participant in the Kyoto process, has said it wants to join and have a target, but so far one has not been specified. Monaco subsequently ratified in May 2006.

The Canadian Liberal government was among the first to ratify the protocol on December 17, 2002, but in January 2006, when the Conservative government was elected in Canada, it was openly hostile to the treaty. Although it has not technically repealed the ratification, the government has said it will not meet its Kyoto targets and cut a popular energy program for homes. In response to the public outcry, the government produced a Canadian Climate Change plan that will allow emissions to continue to rise.

The targets cover emissions of the six main greenhouse gases:
Carbon dioxide (CO_2)
Methane (CH_4)
Nitrous oxide (N_2O)
Hydrofluorocarbons (HFCs)
Perfluorocarbons (PFCs)
Sulphur hexafluoride (SF_6)

These are commonly called "the basket of greenhouse gases." The last three are actually families of complex gases and nearly all of them are produced by industrial processes. They can mostly be captured and destroyed, reused, or recycled to avoid adding to global warming.

Carbon credits
As well as straight reductions of these gases, there are a number of credits nations can earn toward their targets. These include taking action in other countries to reduce their emissions. The idea is that since the atmosphere is a shared resource, a country taking action anywhere to curb emissions is still serving the common good. There are a variety of ways of doing this, all carefully policed by various convention bodies. In addition, there is a system of carbon trading where a country that has exceeded its target can sell the tons of carbon saved to another country that has failed to do so. This system has been adopted internally inside the EU and internationally. Some believe it will allow Russia and some eastern European countries to profit at the expense of countries that are failing to reach their targets. Others accept that even if that is so, carbon trading allows the industrialized countries to reach a collective goal to reduce emissions and establishes a system that will allow more stringent targets to be set later on.

One important point about the convention is that it has no end date; The idea is to continue adding new targets and timetables until a safe climate has been achieved. Each year countries have to report progress toward reaching their targets and how they propose to make up for any deficiencies. There are fears that many countries are not on course to reach targets, and in March 2006 a compliance committee was set up under the Kyoto Protocol to address the problem. It can make suggestions and embarrass countries that are failing to keep to their legally binding targets.

Penalties for not reaching the target, apart from diplomatic embarrassment, come after 2012. By that time a second set of commitments up to 2020 will be negotiated. Those countries that have failed to reach their first target have it added to the second period, plus a 30% penalty for the shortfall.

Mad, Bad, or Greedy?

"I have long believed that claims of a consensus that man is causing global warming is the greatest hoax ever perpetrated on the American people."

—James Inhofe, Oklahoma Republican senator and chairman of the U.S. Senate Committee on Environment and Public Works, May 2006

Contrarians, or climate skeptics, as they like to call themselves, spelled with a "k" here since they are mostly Americans, are taken seriously by the media.

They are a handful of people who crop up hundreds of times in press clippings and on the air, casting doubt on the scientific evidence of global warming and its potential dangers. Some are no doubt sincere and independent in their views. Many of them are paid by fossil-fuel interests and freely admit it, although it seems to do nothing to damage their credibility. Their apparent aim is to spread confusion as to whether climate change is real at all or, if it is happening, to question whether man has any responsibility. Sunspots, a change in the Earth's axis, general unexplained natural variability, and references to warm periods in the past are all used to discredit the mainstream science, even though these factors have already been taken into account by mainstream climate scientists. The second part of the message is to say that action on climate change will cost vast sums of money, wreck the American and world economy, damage international trade, and cost millions of people their jobs.

These skeptics are outnumbered 1,000 to one by scientists who usually have better and more relevant qualifications, along with impeccable credentials. These are people who have increasingly serious concerns about the fate of the planet and humanity, but who stick to properly reviewed scientific data to make their case. They find themselves criticized on radio and television by people funded to do so by the fossil-fuel industry, who claim man-made climate change is unproved. Even some right-wing newspapers encourage their journalists to write thousands of words quoting people with shaky qualifications, often in different fields. This is all in support of their editorial position that global warming is not a problem and should not interfere with the interests of businesses of all kinds, which should be allowed to continue to pollute as usual.

The point is that these skeptics are not operating alone. There is a network of anticlimate science think tanks, lobby groups, and non-government organizations passing themselves off as concerned for the public good. A key to such groups is the names that crop up in their literature and are quoted with approval by them. These include experts like Dr. Robert Balling, Dr. Fred Singer, Dr. Patrick Michaels, Dr. Richard S. Lindzen, Philip Stott, Gerhard Gerlich, Willie Soon, and Sallie Louise Baliunas, who received an award for her "devastating critique of the global warming hoax." One of them, Dr. Lindzen, has produced good papers on water vapor and its effect on climate change that have challenged existing computer models. But, as usual with the skeptic lobby, his name has been mixed in with less-credible and less-qualified propagandists. All of them are climate skeptics. Most derive considerable income as a result. If they are sincere, they are closing their eyes to the weight of evidence. Are they genuinely mistaken, mad, bad, or greedy?

Being against accepted climate science is a good living. Sixty organizations have been identified in the United States alone as taking money from the fossil-fuel industry to discredit global warming science and to campaign against any action designed to curb the burning of oil and coal. There are more organizations in Europe, Canada, and Australia, most of them sponsored directly by industry. Exxon, the world's biggest oil company, has since 1998 given over $18 million to groups opposing the Kyoto Protocol. But this is not a new phenomenon. The fossil-fuel industry realized early that fears about climate change were a potential threat to its business. The message has altered over the years and become more subtle, but during that time, the key messengers have worked their way to the centers of power.

Previous spread: An oil lake lies near a blazing pipeline outside the southern Iraqi city of Basra, March 29, 2003. This was a familiar scene across all of Kuwait a decade earlier when the retreating Iraqi army destroyed the oil wells and let millions of gallons of oil flow into the desert. Then they deliberately set fire to them. In this case the Basra South Oil refinery, which can produce 140,000 barrels per day, was mostly undamaged during the American-led invasion, and workers said they were ready to restart production as soon as electricity and crude supplies were resumed.

Opposite, top: General Motors corporate headquarters in Detroit, Michigan, July 2005. Big consumer incentives were needed to offload gas-guzzling models as oil prices stayed high because of damage inflicted by Hurricane Katrina, continuing problems in Iraq, and anti-American sentiments in Central America.

Opposite, bottom: The Peabody Black Mesa Mine in 2001. The giant coal-mining company pumps water from an aquifer beneath Hopi and Navajo lands for a slurry pipeline, which both tribes claim is sapping their water supply and contributing to a drought. Peabody has been at the forefront of a successful 16-year campaign to prevent the United States from taking any action to curb greenhouse gases and cut the use of fossil fuels.

"The position of the United States is hard to accept given the danger the planet is facing."

—Nelly Olin, the French environment minister, at Montreal, November 2005

Bizarre though it may seem now, the fossil-fuel lobby's first significant campaign was to promote the need for far more carbon dioxide in the atmosphere. Fred Palmer, then head of the coal company Western Fuels, now subsumed into Peabody Energy, the world's biggest coal producer, conducted a campaign at the beginning of the 1990s to triple the amount of carbon dioxide in the atmosphere. He made a film, *The Greening of Planet Earth,* which put forward the idea that crop yields would be boosted between 30% and 60% if carbon dioxide were pumped into the atmosphere. The Earth would enjoy an eternal summer, and world hunger would be abolished. Among those who saw the film just before the Earth Summit in 1992 were President George Bush, Sr., his chief of staff, John Sununu, and Bush's senior energy secretary, James Watkins, who mentioned the movie as a credible source in interviews about climate change. Palmer is now a vice president at Peabody Energy, a company that heavily contributed to the presidential campaigns of George W. Bush. The organization he formed still exists as the Greening Earth Society.

As climate talks continued during the 1990s, the funding of skeptic scientists and lobbyists began in earnest. It has long been a feature of U.N. negotiations that companies cannot have a direct voice, but non-government organizations and pressure groups can have accreditation to lobby delegates with their point of view. Journalists were familiar with the sometimes slick photo opportunities of Greenpeace and earnest youngsters from Friends of the Earth and the World Wildlife Fund (WWF) hovering on the edge of press conferences providing worthy quotes, but they were unprepared for the surge of men in suits. One of the first groups to be identified as "the other side of the argument" was the Global Climate Coalition, supported by the fossil-fuel lobby, car manufacturers, and other heavy industry of the old energy-intensive type. About 50 companies that thought they might suffer a loss of sales if there were restrictions on carbon dioxide emissions paid to belong. Note the name. It is sufficiently misleading to make the casual observer believe they might care about the climate. Their slogan was "Growth in a global environment." In the mid-1990s any of the numerous meetings involving climate science or politics might feature up to two dozen such organizations. They would turn up anywhere in the world. Each one came with lobbyists looking for delegates to convince of their case or, as it seemed to me, at least confuse everyone as much as possible.

There are now even more. Among them are the Alliance for Climate Strategies, which includes the American Petroleum Institute among its members, and the Cooler Heads Coalition that was formed "to dispel the myths of global warming by exposing flawed economic, scientific, and risk analysis." Misleading names are an important part of the game. The Center for the Study of Carbon Dioxide and Global Change promotes junk science as an "antidote" to mainstream government sources.

The most famous lobbyist of them all, and the most effective, was Donald Pearlman. He was a partner in the law firm of Patton, Boggs, and Blow in Washington. He was frequently seen in congressional corridors. For years he never missed a meeting, however insignificant, on either climate science or politics. He was first exposed in 1995 by the German magazine *Der Spiegel,* who called him the high priest of the carbon club. Pearlman was recorded as attending every one of the 20 meetings leading to the first Conference of the Parties of the Climate Change Convention in Berlin in 1995. He saw himself as a "preserver of

(continued on page 100)

Above: This is where the Texas love affair with oil began. Under the drill, the noise started as a rumbling. Then it turned into a deafening roar as thick black liquid exploded 150 ft. (50 m) into the air, tossing heavy sections of pipe around like a handful of matchsticks. The discovery 100 years ago at Spindletop Hill, near Beaumont, marked the birth of the modern oil industry in Texas. It ushered in an era of prosperity, which the state is eager to continue.

"Politicians from other countries pretend that with enough patience and understanding, they will bring the world's largest emitter of carbon along with them to address this unprecedented global threat. They will not. In the U.S., the White House has become the East Coast branch of Exxon Mobil and Peabody Coal, and climate change has become the preeminent case study of contamination of our politics by money. What is missing from all these discussions is the sense of desperation and helplessness which is shared by all of us who are shaken by the effect of each new impact of our inflamed atmosphere."

—Ross Gelbspan, American author and Pulitzer prizewinner, during the Montreal climate talks, December 8, 2005

Top: Exxon Corporation Chairman Lee Raymond (left) shakes hands with Mobil Corporation chairman Lucio Noto at a press conference in New York on December 1,1998. The two discussed Exxon's acquisition of Mobil in an $80.1 billion stock transaction to create the world's largest oil company. The new company continued to question the existence of climate change, funding skeptical scientists and pressure groups while rejecting involvement in the renewables industry.

Bottom: President George W. Bush delivers his State of the Union Address while Vice President Dick Cheney (left) and House Speaker Dennis Hastert (right) look on in Washington on January 31, 2006. It was the first time the president had expressed concern about dependence on imported oil and the need to take steps to reduce it.

On this page: Thousands of old tankers, frequently crewed by poorly trained seamen from developing countries, deliver crude oil to their markets. Many travel thousands of miles from the Middle East to oil refineries in America, Europe, and Japan. At the top the MV *Ulla,* carrying toxic ash from three thermal power plants in Spain in an attempt to export it to another country, sank outside Iskenderen, Turkey, in 2004 causing a major pollution problem, even after Greenpeace called on Spain to take the waste back in 2000. At the bottom the French navy's ship *L'Ailette,* equipped with vacuum pumps, approaches an oil slick threatening the French Atlantic seaboard with an ecological disaster in 1999. The oil spill was from the Maltese-registered tanker *Erika,* which broke in two in heavy seas in the Bay of Biscay.

Above: In 2000 the rescue ship *Smit Langkawi* (front) tows the damaged oil tanker *Natuna Sea* through thick sludge in waters off Singapore. More than 7,000 tons of crude oil flowed into waters off Singapore and Indonesia from this Panama-registered ship after the vessel ran aground near one of the world's busiest shipping lanes.

> "Saving our civilization is not a spectator sport."
>
> —Lester Brown, Plan B 2.0

Above: When the oil from sunken tankers reaches the shore, the cleanup is very messy, extremely difficult, and often dangerous. Here, volunteers clean up fuel oil on Muxia's A Pedrina beach in Galicia in northwest Spain on December 6, 2002, after thick fuel oil was oozing from the sunken tanker *Prestige* off the coast and drifted onto the beaches.

Top: Physically carrying away the oil in buckets is frequently the best way of cleaning up, because detergents often make pollution worse. Two stained volunteers carry a bucket filled with fuel oil spilled from the oil tanker *Prestige.* The aging single-hulled tanker foundered off the coast of Galicia in November 2002 with 77,000 tons of heavy fuel oil on board, causing Spain's worst-ever ecological disaster, contaminating hundreds of miles of coast and putting thousands of fishermen out of work.

Bottom: Thousands of birds died like this one, unable to escape from the crude oil in the sea that leaked from the *Prestige.* Fish and wildlife were affected along hundreds of miles of coast.

> "This is a rogue administration, out of step with the rest of the world and much of America, representing just a small number of powerful industrial interests."
>
> —Jonathon Porritt, chairman of the UK Sustainable Development Commission, December 9, 2005

the American way of life" and called his organization the Climate Council. He operated out of the same building as the Global Climate Coalition but never revealed who his clients were, although his law firm represented, among 1,500 clients, Exxon, Shell, Texaco, and DuPont. Pearlman was frequently seen in the company of delegates from the Gulf states, who wore the flowing robes of the desert whether they were from Kuwait or Saudi Arabia. He was said to be more familiar than anyone alive with the 1,000 documents on climate change produced before the conference. He had argued about every line. He passed endless written notes to the men in robes, which observers claimed frequently turned into obstructive demands from the floor of the conference hall.

Unlike many lobbyists, Pearlman did not seem concerned with the press. In his early years he would simply say "no comment" to any question about whom he represented. After the *Spiegel* article, he refused to be drawn out on any subject, including his aims or beliefs. In later years, he simply turned his back. He died in 2004 after single-handedly holding back progress on the science and politics of climate change for years. He did not live to see the Kyoto Protocol come into force. Most of his best work was done during the Clinton years, when there was some American impetus to doing something about climate change. He managed to stall progress long enough for an oil man to gain the presidency.

When George W. Bush was elected, the fossil-fuel industry became closely connected to the White House. It was not just that Bush himself was an oil man. Fossil fuel companies partly funded his campaigns. Some of the White House staff were also former or actual lobbyists.

Harlan Watson, the U.S. chief negotiator on climate, had been a friend of Pearlman's. It was Watson, as a Republican congressional aide in the early 1990s, who urged the coal industry to hire Pearlman as a lobbyist. It was Exxon Mobil that suggested Watson be added to the administration's climate team. Perhaps just as significant, Paula Dobriansky, Bush's undersecretary of state for global affairs since 2001, met Pearlman shortly after Bush pulled out of the Kyoto agreement. She said the purpose of the meeting was to "solicit [his] views as part of our dialogue with friends and allies." She also thanked Exxon for their involvement in formulating climate policy.

In June 2005 the extent of infiltration of the White House became clear when the *New York Times* revealed that Philip Cooney, a Bush aide and former oil lobbyist with the American Petroleum Institute, had doctored reports on climate change over a four-year period. Climate research from such internationally trusted and august bodies as the National Academy of Sciences, the National Ocean and Atmospheric Administration, and NASA were altered and in some cases passages omitted, after they had been peer reviewed and adopted officially. Shortly after he was exposed, he resigned and was immediately hired by Exxon Mobil.

Although the fossil-fuel lobby has been remarkably successful in obstructing progress on dealing with climate change, the traffic has not all been one way. The Global Climate Coalition began to lose members rapidly as the 1990s drew on. The first oil company to withdraw was British Petroleum, which realized that the coalition's purpose to "cast doubt on the theory of global

---

Previous spread: It is not just old tankers that cause oil pollution, nor is it only wildlife and fishermen that suffer; the tourist trade can also be devastated. This is an aerial view of an oil spill near Rio de Janeiro in January 2000. Brazil's state oil giant Petrobras said an underwater pipe had ruptured, spewing about 500 tons of oil, which began washing up on beaches near Rio de Janeiro city. The Petrobras president, Henri Philippe Reichstul, called the rupture a "worrying accident," which he estimated caused a moderate oil slick 2–3 miles (3–5 km) long.

Opposite: Exxon became the target of protests across the world for its stance on global warming. Exxon claims that a boycott of its products launched worldwide has had no effect on sales. With a company that measures its daily profits in the millions of dollars, it is hard to tell. Lee Raymond, the chairman and chief executive, became a symbol for all that was wrong in the oil industry for environmental campaigners. He was dubbed the Darth Vader of global warming. There was an outcry in June 2006 when he retired after 12 years as the head of the world's largest oil company with a retirement package of $400 million. During his leadership the company pumped an estimated 6 billion tons of carbon into the atmosphere and contributed to George W. Bush's election campaign and many groups that deny global warming across the United States and elsewhere.

warming" might not fit in with its new solar business. The coalition began to collapse in March 2000 when Texaco defected, but other members, notably Exxon Mobil and General Motors, remain active in discrediting climate change, and the website still remains full of misinformation.

Of course it would be wrong to ignore the fact that there have been lobbyists from the other side, who have worked extremely hard to persuade leaders from every country to take more action on climate change. They are not averse to infiltration, either. Some of the European countries have included members of the Friends of the Earth and The World Wildlife Fund (WWF) as part of their government delegations, perhaps helping to counterbalance the oil men in the giant U.S. camp. There are literally dozens of non-governmental organizations, from one-man bands to large groups with 20 or more lobbyists. The Greenpeace lobbying organization at any talks is usually larger than 90% of developing-country government delegations. Each group has its own point of view, and some, like Greenpeace, would refuse to join government delegations to avoid compromising their position of being able to criticize anyone.

Outside the conference halls, too, the insurance industry has always lobbied hard on its own account and backed fringe meetings and scientific presentations. Dire warnings on the escalating insurance claims because of weather-related incidents form the basis for reports about the uncertain future of the industry's own member companies, as well as the planet, if global warming is allowed to continue unchecked. So the skeptics have not had it all their own way. The disadvantages and the economic ruin the world faces because of climate change have been assessed and updated at meeting after meeting and made available to journalists and politicians.

Top: Its sales badly dented by Japanese automakers producing small-engine cars and hybrid vehicles, General Motors, finally responded in 2006 by unveiling the new Saturn VUE SUV hybrid. The vehicle, which costs around $25,000, still uses more gasoline than the average European car.

Bottom left: Despite the continuing rise in oil prices and publicity about global warming, U.S. automakers still do not seem to have gotten the message. Here, the 2005 Jeep Grand Cherokees sit on the assembly line during their production-launch celebration at the assembly plant in Detroit.

Bottom right: In an attempt to cut costs, Ford has switched to robots to build its cars. Approximately 380 robots build the body of the 2005 Mustang. Subsequently, Ford has announced that it will compete in the smaller, greener car market with new models to try to recover some of its former worldwide dominance in the family market.

Opposite top: Many conflicts around the world are described as battles for freedom and democracy or for the overthrow of dictators. But many of the most bitter conflicts are over who controls valuable natural resources, and this is often oil. Perhaps it is a coincidence that as this Russian soldier lights a cigarette in front of a tank, there is an oil pipeline burning in the distance. He is on the outskirts of the Chechen capital Grozny in 2000.

"I think the environment should be put
in the category of our national security.
Defense of our resources is just as
important as defense abroad. Otherwise
what is there to defend?"

—Robert Redford, Yosemite National Park
dedication, 1985

So it is possible to argue that this is democracy at work. It is a valid argument. After all, both the insurance industry and the fossil-fuel lobby are protecting their own interests, and to a large extent the environmental organizations need to be campaigning to garner new members. However, when thousands of lives are being lost because of climate change, which may soon be in the millions, should more be expected from lobbyists?

The fossil-fuel industry has been party to misinformation, obstruction, and corruption. Warnings from the UK's Royal Society that this was still happening in the run-up to the release of the latest international scientific findings on climate change in 2007 showed that the battle to discredit the work of the Intergovernmental Panel on Climate Change (IPCC) still continued. Even in the hours before the IPCC report on impacts across the world was released on April 6, 2007, government representatives from the United States, China, Saudi Arabia, and Russia succeeded in watering down the final report to make the science appear less certain than it is. For example, a graph showing that billions of people would be at risk of coastal flooding by 2080 was changed to read millions. Some scientists walked out in protest. It is hard to understand a motive, except greed, for all this time and money spent to discredit science and damage the political process. Some of the scientists involved with this lobby are no doubt sincere; there are, after all, still some legitimate uncertainties about the speed and extent of climate change. But the main motive of those that sponsor, promote, and employ them, the corporations and the oil-rich countries, is to protect their interests. In other words it is about protecting profits at the expense of many lives.

Bottom: In the Niger delta in Nigeria, another of the world's oil hot spots, local people whose land has been ruined by pollution from oil exploitation have been involved in a long-running struggle with oil companies and the central government. Here in 2000 a fire continues to burn along a ruptured pipeline in Oviri court in Adedje in the Niger delta after an explosion on the pipeline claimed the lives of 250 people. It was the latest in a series of incidents in which people, who were believed to be illegally tapping into the pipelines, had been burned to death.

Following spread: Whatever the motives were for invading Iraq, restored oil supplies were supposed to launch the newly occupied Iraq into a new era of peace and prosperity. The reality more than a year after the United States–led invasion was a reduction in oil exports after the sabotage of the country's main internal oil pipeline. The instability has led to increases in the price of oil and the occupation has been a severe drain on the U.S. economy. Policemen guard the burning pipeline near the city of Kerbala, around 70 miles (110 km) south of the Iraqi capital Baghdad on February 23, 2004, as the country gradually descended into chaos.

Even Termites Are to Blame

Previous spread: Sunset over a garbage dump in Guiyu, one of the most polluted districts in China. It is the final resting place of unwanted hi-tech waste from Europe and America. Electronic equipment, including televisions, computers, and cell phones, are the fastest-growing toxic-waste stream in the world. Although the export of such waste is banned under the Basel Convention, which all industrialized countries except the United States have ratified, much of it is dumped in China and other developing countries because it is claimed to be for recycling or reuse. Where electronic goods are dismantled by hand to obtain some of the precious metals inside, the workers and communities involved are exposed to serious environmental and health problems from metals such as mercury and lead.

Above: Pollution mishaps are frequent in the middle of one of the world's most troubled oil fields. This spurt of oil sprang from a supposedly sealed wellhead belonging to the Shell Oil facility at Kegbara Dere in Ogoniland, in Nigeria's volatile Niger Delta, in March 2006. Shell, which has been under pressure to improve conditions in the delta, later admitted it had detected an oil spill at a dormant wellhead in the Ogoni area.

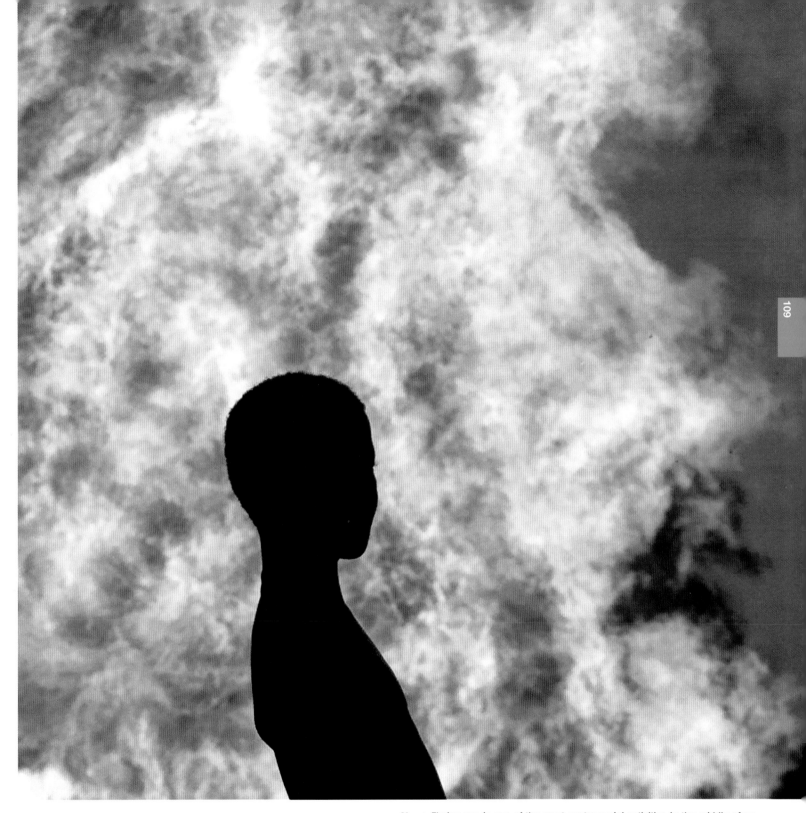

Above: Flaring gas is one of the most controversial activities in the middle of an oil field, because it is a relatively clean fossil fuel that could be captured and sold rather than burned. A Nigerian child is silhouetted against a gas flare at the Shell Utorogu facility in Nigeria's southwest delta in 2004. This was taken on the same day that Nigerian troops dislodged hundreds of protestors. They had blocked access to Shell's facilities in the delta region and threatened to shut in 100,000 barrels of crude-oil output.

110

There is no doubt that the temperature at the Earth's surface is rising and rapid changes are taking place around the planet as a result. Three massive reports running into thousands of pages, with experts from all over the world considering every aspect of the Earth's ecosystems, have confirmed this, along with one of the major causes of the changes—the burning of fossil fuels.

It has been the biggest scientific collaboration in human history, and it has led to the agreement of thousands of scientists that the planet is in a dangerous state. Yet somehow this process has still left room for some to claim there is uncertainty and doubt.

As stated in the previous chapter, the work of the Intergovernmental Panel on Climate Change (IPCC), which has produced these four convincing consensus reports over a 21-year period, has been obstructed and interfered with at every opportunity to try and make the wording as bland as possible. Even so, to any objective reader the reports make extremely gloomy reading. The contrarians, however, have achieved their purpose. Over the years enough ifs, buts, maybes, and uncertainties have been introduced into the text for cynical supporters of the fossil-fuel lobby to question many of the scientific findings.

However, the 2007 reports changed that, especially the one released in April that described the impact of global warming in different regions. It was striking that the actual measurements and observations of real events on the ground accorded exactly with what scientists had predicted would happen in previous reports. It gave great confidence that the computer predictions for the future were also realistic.

This was despite the fact that the United States, Saudi Arabia, China, and Russia continued to intervene in the hours before the press conferences to launch the summaries of the reports. The political representative sometimes successfully toned down the language to the anger of the scientists who had originally written the reports and were to present the findings. Among the key findings that survived this watering-down process was that hundreds

of millions of people would have less access to fresh water, particularly in the already stressed drought-prone and semiarid regions. Approximately 30% of all the world's species would be at greater risk of extinction, corals would be damaged by the increased heat of the sea in the tropics, and many reefs would die completely. Fish species would shift in response to warmer water, and more carbon dioxide would be released into the atmosphere because of the damage climate change would do to forests and other natural systems.

Food production would also be hit in many regions, although some countries in the colder regions of the Northern Hemisphere might have short-term gains because of a warmer climate. Perhaps the worst news was for coastal dwellers and health. Millions more people would experience coastal flooding each year and be forced to retreat inland, and for almost everyone the warmer weather would bring more infectious diseases and diarrheal infections. Extreme weather events of all sorts would increase, especially tropical storms and flooding.

Regional predictions that had not been given before showed that many areas could suffer severe impacts and some could gain. For example, North America could experience some gains in cereal production from rain-fed agriculture—particularly in Canada. However, the United States could also experience difficulties in the South where crops are at the "warm end of their suitable range" or rely heavily on irrigation. The western mountains, particularly in California, would lose snowpack, experience winter flooding, and reduced summer flows, making competition for already scarce resources worse. Pests, diseases, and fire could be expected to have adverse effects on forests, with larger areas burned. In the cities

Above: Scavenging in garbage is one of the last resorts of the very poor. In this photo Filipinos hike to the rubbish mountain called the Promised Land in Manila. It is dangerous work. Only days before this picture was taken in 2000, an avalanche of garbage demolished more than 200 shanties and 137 bodies were removed by rescue workers.

Top: This Indonesian scavenger has taken to a raft in a garbage-laden canal in Jakarta on World Environment Day in June 2003. Many of Indonesia's impoverished live on the banks of the city's canals, which overflow during the rainy season.

Bottom: In South America the garbage problem can also get out of hand. After flooding in 2003 caused 50,000 to flee their homes in Santa Fe, Argentina, these children were left to play among piles of garbage heaped outside homes.

Above: An old man strains to pedal his tricycle into a sandstorm that has turned the sky amber and greatly reduced visibility in Beijing, March 20, 2002. Extensive deforestation and desertification in northern China have fueled dust storms. Nearly 1 million tons of Gobi Desert sand blows into Beijing each year.

> "Climate change will continue to exert profound influence on the ecological environment as well as the social-economic system in China, and such influences are mainly negative."
>
> —Liu Jiang, vice chairman, National Development and Reform Commission of China, February 2005

Rising Temperature Forecast

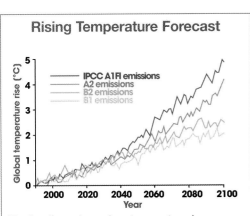

- IPCC A1FI emissions
- A2 emissions
- B2 emissions
- B1 emissions

The four lines above show temperature rises over the next century. The rise depends on the quantity of greenhouse gases emitted in the first years of this century. Temperatures over land are expected to increase about twice as rapidly as temperatures over the ocean. The lowest line represents the increases forecast if emissions are reduced by 60% below 1990 levels. The top line is a continuation of current "business as usual." The lines in between represent some of the factors, such as adoption of new technologies or oil running out, which would affect emissions. Scientists say both of these scenarios are not nearly radical enough.

SOURCE: IPCC/Hadley Centre for Climate Prediction and Research

heat waves would kill more people, particularly the elderly. Coastal regions would be hit by more storms and sea-level rise.

Compared to the impact on small island states and polar regions, North America escapes lightly from climate change. Small low-lying islands would lose beaches, coral reefs, and some fisheries, and it would therefore have an adverse effect on their main income from tourism. Storm surges, erosion, and other coastal hazards might make some islands uninhabitable, particularly where saline intrusion threatens the water supply. The report says the polar region's human communities—adapted to living on the ice and on frozen ground—will soon find their entire lifestyle threatened, and special measures would be needed to help them to adapt and survive. Many polar species, migratory birds, mammals, and predators such as polar bears would struggle to survive.

Latin America would lose a lot of forest, including parts of the Amazon that was expected to turn into dry grassland. The loss of glaciers in the high mountains would cut off summer water supply for irrigation and energy generation, and human consumption would force people to abandon their farms and their homes. There would be considerable loss of species both on land and at sea.

Europe, too, would expect adverse effects—particularly in the south, where warmer, drier conditions would damage agriculture, causing desert conditions in parts of Spain, Italy, and Greece. The Alps would lose glaciers and most of the skiing industry. Eastern Europe would also be drier and suffer heat waves. In the north there would be fewer heating costs in the winter, better forest growth, and higher agricultural yields, but winter floods and

ground instability because of permafrost loss would cause problems.

Asia and Africa—the two continents that have done the least to contribute to climate change—would suffer most. Glacier melt in the Himalayas would increase flooding, and rock avalanches and destabilized slopes would affect water resources. Decreased river flows would occur as the glaciers recede. More than 1 billion people in India, China, Pakistan, and Bangladesh are expected to be short of fresh water by 2050. Coastal regions, particularly the mega deltas of Asia and Africa, would suffer floods, coastal erosion, and storm surges. Corals and mangroves would be lost. The new report also confirmed that Africa would be the continent most vulnerable to climate change. By 2020 up to 250 million more people will be short of water, and food shortages will grow dramatically because of climate change. Even the fisheries in large lakes will be badly affected by rising water temperatures, a situation made worse by overfishing.

These disturbing predictions are for a much shorter timescale than previous reports, and they make it clear that some of the impacts are already happening. One of the points made by scientists and environmental groups alike is that these predictions are conservative and often already out of date. The rigorous process of peer reviewing all of the scientific findings before they are even considered for inclusion in the IPCC reports means that much of the science included in the 2007 report was already two to five years old. Since the report was compiled, there have been many more observations showing that climate change is moving faster than this devastating report predicted—more ice is melting, and extreme temperatures, droughts, floods, and storms are more frequent. Climate change is gathering pace. Part of the problem in understanding climate change is that nearly every statistic is an average. What is frequently misunderstood is that temperature rise over the land is usually twice as fast as over the sea, and in the Arctic regions it is three times as fast as in the tropics. This has severe effects on ice melt and on sea-level rise as can be seen in the next chapter. It could lead to the release of massive stores of greenhouse gases locked up in the tundra. Western Siberia, for example, is heating up very fast. The temperature has risen 3°C in the past 40 years.

This and some of the other problems are discussed in greater detail later on, but it is always important to understand the inter-relationship between all of these problems. Each one tends to make the next one worse, or to put it another way, the changes happen more rapidly and with greater extremes. Ultimately they affect the ability of the human population to survive, especially the current numbers, let alone the extra billions expected to be born between now and the middle of the century. Melting ice and tundra leads directly to the next problem, and in the longer term the most intractable problem for the human race: sea-level rise. Quite simply, the land area on which we live and grow crops is shrinking and will go on shrinking for 1,000 years because of the warming we have already committed.

Not only are the glaciers, Greenland, and other ice caps melting into the sea, the oceans are expanding as they warm. This and other related effects of warm seas are becoming apparent. For example, fish species, which have adapted to the ocean as it has been for thousands of years, are very sensitive to temperature changes. The temperature of the water affects their ability to

On this page: Everyone in China has been called upon to take part in the country's vast tree-planting program. Partly to blame for the trees' destruction in the mountains are the disastrous floods in recent years and evidence of rapidly spreading deserts. Pensioners, soldiers, and children all plant trees. At the top soldiers dig the holes while schoolchildren carry the saplings to be planted. At the bottom a child rests between planting sessions. Both photos were taken 45 miles (70 km) from Beijing, where belts of trees are being planted to protect the capital.

Opposite: Where new forests have apparently been successfully replanted, the Chinese are taking no chances. This plane is on an air-seeding mission in Lushi, central China's Henan Province. In 2005 the local government conducted air seeding over 16,000 acres (6,500 hectares) and claims to have reduced the area of desert under their control. However, in China as a whole, the total land area affected by deserts is said to increase by 800 sq. miles (2,000 sq km) a year.

breed; just as important, the tiny creatures they feed on, such as plankton, are even more sensitive to warmth and either disappear or move on. Oceans are also affected by increased acidity, as mentioned earlier. This is another potential disaster for the marine food chain and one of mankind's major sources of protein.

Although the droughts in southern Europe in the first years of this century and the extreme heat wave of 2003 have alerted many in this otherwise green continent to the potential spread of deserts, it is mostly in the developing world that a loss of croplands has been causing alarm. One of the theoretical predictions of long-term climate change is that it will start to rain again in the Sahara, but currently the world's largest desert is still expanding. It seems intent on crossing the Mediterranean to Spain, Italy, and Greece. China is one of those countries where dust storms and desertification have been ringing alarm bells for years. A massive program of tree planting aims to stem the tide.

Although natural forces change the boundaries of deserts and in some cases dry areas can be reclaimed by irrigation and careful planting, man is also responsible for creating deserts and enlarging them. Cutting down forests, overgrazing dry land, and poor irrigation practices, which cause salt to build up in the soil, have massively increased the area of desert in China. Climate change adds to these pressures and makes rehabilitating deserts more difficult, pushing even greater numbers of people to seek employment in the cities because they can no longer feed themselves on the land. Population pressure is one of the taboo subjects in the environment debate. But how do you continue to feed this ever increasing population in a warming world? It was one of the great debates at the Earth Summit in 1992, when the Roman Catholic Church threatened to withdraw support if birth control was discussed.

Those issues will be discussed again as we illustrate how the world is changing, but it is also important to emphasize the relationship of climate change to other environmental problems; the role of air pollution in its various forms, for example. The public, and even sometimes the scientists, have been slow to understand the links. Acid rain is a good example of this interlinking. In the 1970s in Europe, there was a long battle to understand and then, at a political level, take action to combat acid rain. All over northern Europe life in lakes was dying, even though they looked crystal clear, and pine trees were losing their needles. Whole forests were denuded. The culprit was the low-quality coal burned in power stations. The coal produced noxious fumes in the form of sulphur dioxide, which in dry weather was deposited directly onto the ground but often mixed with the clouds, producing a weak solution of sulphuric acid. Smoke from British power stations drifted hundreds of miles to the northeast and fell as acid rain in southern Scandinavia.

The pattern was being repeated all over Europe and the world. The damage was not just to fish and trees. Acid soil changed the types of plants able to survive in certain areas and damaged buildings; for example, gargoyles on cathedrals carved of limestone literally dissolved. The problem of acid rain is solvable, and rich communities such as those in Europe and North America set about reducing sulphur emissions by filtering smoke before it was released. This had a surprising side effect. Unlike carbon dioxide, which effectively remains in the atmosphere for approximately 100 years, sulphur dioxide lasts only a few days or at most weeks.

Opposite: Acid rain has damaged forests and stone monuments across large areas of Europe and is an increasing problem where coal is the main energy provider in the developing world. The acid comes from sulphur in coal and oxides of nitrogen, producing weak solutions of acid when smoke mixes with the clouds. An extreme case is the dead spruce trees in the Ore mountains in the Czech Republic.

Above: More typical of acid rain damage is the dissolving of this statue in Krakow, Poland. Many old buildings were built with limestone and cemented together with limestone cement. Acid rain dissolves both stone and mortar, resulting in irreparable damage. Statues made of marble, a form of limestone, are also susceptible to acid rain. The Acropolis in Athens was damaged more in the last 25 years than in all of the 2,400 years before. It costs Europe $9 billion per year to replace the corroded stone.

"There may be technology to deal with emissions from the industrial and energy sectors but we have not yet found ways to stop cows and sheep from doing what comes naturally."

—David Parker, New Zealand's minister responsible for climate change issues, speaking at a climate-change conference in New Zealand in 2006, referring to his country's millions of livestock

Above: Rice is a thirsty labor-intensive crop and a staple food across large areas of the world. Climate change is threatening the ability of large areas in China to grow enough rice to feed the population, and everywhere peasants are leaving the land to move to the cities. China expects to use its new economic power to import grain from its neighbors, but for how long? Vietnam is still a net rice exporter, with the grain grown in the traditional manner. Here, a farmer, with ancestors' graves behind and new rice seedlings in front, turns his buffalo as he ploughs to prepare a new paddy field on the outskirts of Hanoi.

Above: In Bangladesh boys follow tradition, washing paddy seedlings at Singair near the capital Dhaka, but the country is also vulnerable to extreme weather, which threatens the food supply. In 2004 2.77 million acres (1.1 million hectares) of crops were submerged by floodwater.

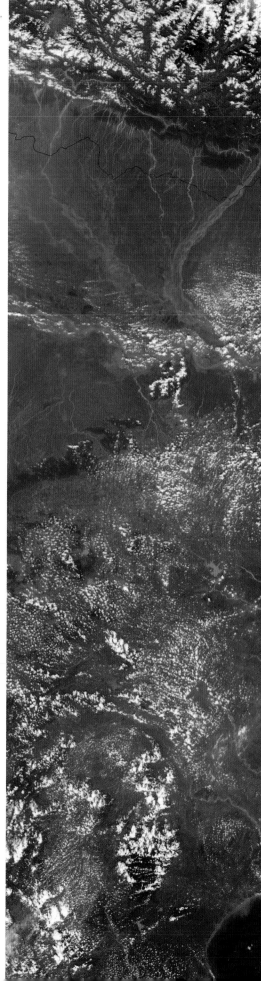

As soon as cleanup began on the smokestacks, the air cleared, and the temperature began to rise. The Mount Pinatubo eruption of 1991 and the resulting 20 million tons of sulphur dioxide emitted into the atmosphere confirmed to scientists studying global warming that pollution had been holding down global temperatures by reflecting back sunlight.

Some scientists now believe that the huge levels of pollution between the 1940s and the 1970s held world temperatures down when they would otherwise have been rising in response to increased carbon dioxide levels. If this is true, it adds to the scariness of current climate change. Some of the NASA pictures from space show massive pollution on either side of the Himalayas. This is from the surging fossil-fuel use in India and China, which has been accelerating over the last 30 years. This, too, must be greatly limiting the rise in temperatures that is already becoming apparent. These two countries are both attempting to tackle the life-threatening pollution in their cities. If they succeed, they will send the temperatures soaring over Asia, according to IPCC scientists.

Another complex interaction between two apparently unrelated environmental problems is the role of ozone depletion. The well-documented "hole" in the ozone layer caused by the release of man-made chemicals into the atmosphere is a problem because it allowed larger quantities of harmful ultraviolet light to reach the Earth. It also altered the balance in the amount of heat retained by the atmosphere. While greenhouse gases heat the lower layer, called the troposphere, ozone depletion caused the stratosphere to cool. While ground stations measured the Earth heating up, satellites, which give more of a global average, showed virtually no warming. It provided a field day for the

Above and right: Water is the biggest potential point of conflict between the three countries that share the great rivers that flow south from the Himalayas—India, Pakistan, and Bangladesh. The water supply is threatened as the glaciers disappear with climate change, and any new scheme to harness the rivers leads to increased tension. Above the Baglihar hydroelectric project on the Chenab River led Pakistan to complain that India was violating the Indus Water Treaty by restricting cross border water flow in order to generate electricity. The picture on the right shows how rivers rise in the Tibetan Plateau (upper left) to nourish the three countries before reaching the Bay of Bengal.

contrarians, throwing doubts on all of the science. Some scientists then concluded that the satellites had been averaging out the warming of one layer with the cooling of another. Discovering precisely what is happening remains elusive, and this issue remains a hotly contested area.

So it seems that although air pollution, acid rain, and the ozone depletion are all serious problems that must be solved because each one has detrimental effects on human health and the environment, their cure will also have a bearing on how fast the Earth heats up. In some heavily polluted areas in Asia, the effect could be large. Although the net result of civilization's attempts to solve these various environmental problems on climate change is still not quantified, there have been more recent attempts to discover how much the climate will warm. As discussed earlier, there is an emerging consensus that anything above 2°C risks disaster. It is also a tall order for many long-lived species, such as trees, to adapt to a rapid rise in temperature. A 1°C rise is equivalent to moving 150 miles south in Europe, and a similar rise would move the tree line 500 ft. (150 m) up a mountain.

In early 2005 a team from Oxford University used the power of 90,000 linked personal computers to run a series of trials to see what effect a doubling of carbon dioxide would have in the atmosphere—approximately 550 parts per million (ppm). As we have already discussed, the temperature that this level of carbon dioxide produces has an important bearing on the political decisions that need to be made to avoid dangerous climate change. This 550-ppm level of carbon dioxide in the air is also one that the IPCC scientists have been using to make their predictions, because based on current trends, that will be the level by 2050. The resultant predicted increases ranged between 1.9°C and 11.2°C, depending on how different processes in the atmosphere were represented in the models. This is not only a wider range than the previous IPCC results, but also a revision upward. Even the lowest figure is perilously close to the 2°C danger threshold. The effect of 11.2°C would be very dramatic.

An important aside here is that throughout, the amount of carbon dioxide in the air has been central to the scientific and political argument. Methane has also been mentioned as a key greenhouse gas. There are other greenhouse gases including chlorofluorocarbons, or CFCs, which are being phased out anyway because they are the major cause of the destruction of the ozone layer. As has been mentioned earlier, there were six greenhouse gases, or groups of gases, mentioned in the Kyoto Protocol, the reduction of all of them counting toward any country's target. Apart from carbon dioxide and methane these are nitrous oxide—otherwise known as laughing gas—and two groups of chemicals called hydrofluorocarbons and perfluorocarbons, which are produced in chemical manufacturing, and sulphur hexafluoride. All of them have a far greater heat-trapping effect in the atmosphere, per kilogram emitted, than carbon dioxide, but they are in such tiny quantities that properly controlled they will not be a problem. Nitrous oxide is released with the production of nylon and is produced naturally in the soil, but excessive fertilizer use is a major human cause. Extensive efforts are being made by industry to control the other industrial gases and to find non-damaging substitutes.

Right: The Ganges is a sacred river used for Hindu ceremonies. It is also a great life-giving supply of water and therefore food to millions of people, all of which are in jeopardy. The river is fed by snowfall and glaciers in the Himalayas and the rain from the monsoons. All of these supplies are changing with the warming climate. Thousands of Hindu worshippers bathe in the holy Ganges in Allahabad, February 8, 2001, during one of the most auspicious days of the Maha Kumbh Mela festival called Maghi Poornima. Tens of millions of Hindus gather over a six-week period to bathe in the holy waters of the Ganges at the festival that is held only once every 12 years.

We Need Ice

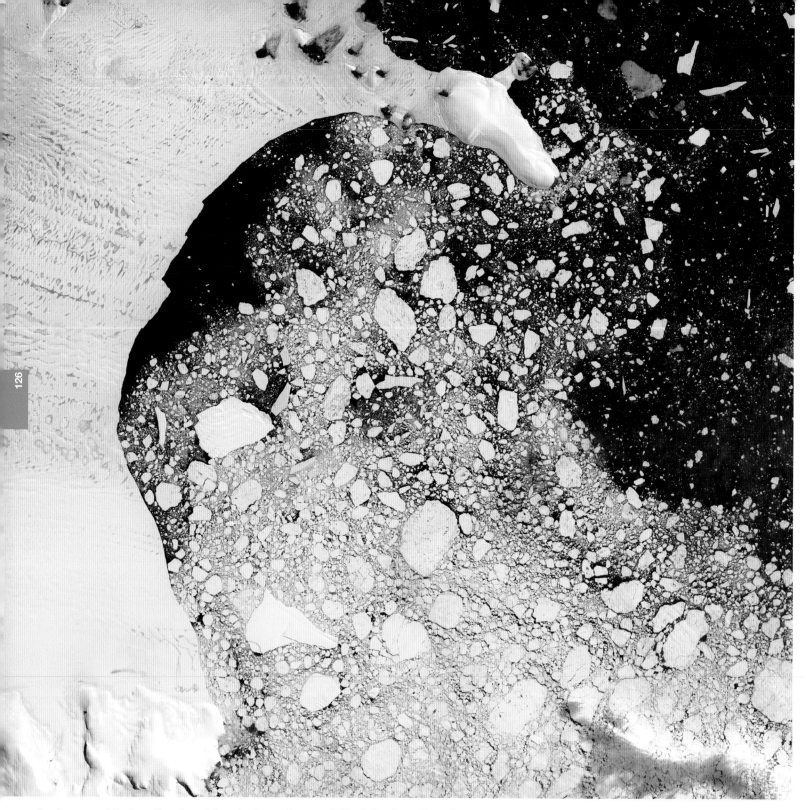

Previous spread: Northern Canada and Greenland, seen from a satellite during the summer of 1999, are still predominantly white as ice and snow continue to cover most of the land and sea. This image will be used in the future to show just how much global warming has affected the ice cover on the roof of the world.

Above: The Larsen Ice Shelf in Antarctica as seen from space. Warmer temperatures over a few months can cause huge areas of apparently stable ice to collapse. Scientists can see pools of meltwater on the surface that fill crevasses and increase the pressure on the shelf, extending the cracks and causing it to disintegrate.

Above: Although this image is not of the same quality as others in the book, this photo, which was taken in 1997 of hikers on top of the Rhône Glacier in the Swiss Alps, is unrepeatable. The Rhône Glacier has retreated so much that these hikers would now be walking on air. The glacier is the source of the Rhône River, which feeds Lake Geneva before running across France to the Mediterranean.

Ice is crucial to the way the Earth's climate works. It is also vital to the survival of many wild creatures and provides the regular water supply for hundreds of millions of people.

The Acceleration of Glacier Melting

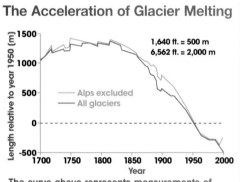

1,640 ft. = 500 m
6,562 ft. = 2,000 m

— Alps excluded
— All glaciers

The curve above represents measurements of glaciers on which there is reliable information; it shows that in the last two centuries these ancient rivers of ice have retreated an average of nearly 6,500 ft. (2,000 m). Between 1900 and 1980, 142 of the 144 glaciers for which adequate data were available decreased in length. The exceptions were in areas where climate change has caused considerable increase in snowfall, replenishing glaciers faster than they melted. Since 1980, with rapidly increasing temperatures, glacier melting has accelerated across the world.

SOURCE: H. Oerlemans, Utrecht University.

But as temperatures rise from Kilimanjaro near the equator in central Africa to Alaska, glaciers are disappearing. In the highest mountains in the Himalayas, around the Greenland ice cap, even at the North Pole and in Antarctica, the accumulated ice of millennia is melting. This is not the only concern. On high mountains and across vast tracts of land in northern forests and treeless tundra, the land is frozen underground and has been since the last Ice Age. More than a quarter of the world's land area is permafrost. Many of these areas are melting, too.

Perhaps the first thought that springs to mind about melting ice is its effect on sea-level rise. The other effects of melting ice are not so obvious but are equally important and, for many, far more devastating. The world's food supply is heavily dependent on the winter buildup of snow on high mountains that is compressed to form ice caps and glaciers. Each spring some of the snow melts, and then during the dry summers the glaciers provide a smaller but reliable and constant replenishment to feed the rivers below.

To understand the increasing impacts of melting ice, it is important to understand the role of ice and snow on the climate. It is calculated that about one-third of the sun's rays are reflected back into space after striking the Earth. The amount of light reflected is not uniform. White ice and snow reflect about 90% of the light and most of the heat. Ocean water, on the other hand, absorbs most of the light and heat. As the world warms and the area of ice and snow declines, the areas around the edge absorb quantities of energy from the sun that were previously sent back into space. This warming, which then causes more warming, is called a positive feedback. This partly accounts for some of the dramatic changes seen in the Arctic in the last 20 years. The temperature in the area has risen as much

as 3˚C. In 2005 experts at the U.S. National Snow and Data Center in Colorado said that in September of that year—the maximum time of ice melt—the extent of Arctic sea ice had dropped 20% below the long-term average. An extra 500,000 sq. miles (1,294,994 sq km), or an area twice the size of Texas, had turned from reflective surface ice to water. This was the fourth successive year that melting had been greater than the average.

The melting ice almost completely cleared the notorious Northwest Passage of ice. This area north of Canada is where many heroic expeditions were lost in past centuries trying to find a shorter sea passage from Europe to Asia. Soon it will be possible to sail unhindered along this route. Already the Northeast Passage across the top of Siberia is open in summer. The satellite measurements also show that in winter the ice never re-forms over such a wide area as it once reached. The spring melt also begins earlier; in 2005 it was 17 days earlier than the 20-year average.

Research carried out by submarines over a 20-year period showed that the thickness of the remaining sea ice had by the spring of 2007 dropped by 40%. This led British researchers to calculate that in summer all sea ice might disappear as early as 2020.

Shortly before the announcement about the loss of ice on the seas, a joint research project by Russian and British scientists revealed a remarkable melting of the permafrost in Siberia. It is the world's largest peat bog. An area the size of France and Germany combined was developing lakes and turning to mud for the first time since it had formed 11,000 years ago. Western Siberia, along with the rest of the Arctic, is heating up faster than anywhere else in the world, recording a 3˚C rise in 40 years.

"It seems to me that the natural world is the greatest source of excitement; the greatest source of visual beauty; the greatest source of intellectual interest. It is the greatest source of so much in life that makes life worth living."

—Sir David Attenborough, in an interview with the BBC, 2006

Above: A shortage of snow in the Alps has caused a crisis at some winter sports resorts. Snow-making machines are used in many places; however, they are very expensive to run and add to climate change because they are powered by fossil fuels. To try to get around this problem, the authorities at Mayrhofen in Austria give skiers and snow-sport enthusiasts a pass that allows them to move to another resort. Snow shortages have also caused similar problems for Canadian resorts, such as Whistler and Banff.

The biggest fear of scientists is that some of the 70 billion tons of methane trapped in the frozen soil will be released by this warming. Since methane is 23 times as potent a global warming gas as carbon dioxide, this is another potential large-scale positive feedback. David Viner, from the University of East Anglia, said at the time: "This is a big deal because you cannot put the permafrost back once it's gone. It will ramp up temperatures even more than our emissions are doing."

The parts of the planet where these changes are taking place are generally the least inhabited. There are Canadian, European, and Russian tribes with settlements well inside the permafrost zone who have been forced to move because their homes have begun to tilt and collapse as the soil has melted. In the south of the permafrost regions, the melting allows forests to migrate north. In Greenland, for example, farmers can grow potatoes where it was too cold before. Farther north, however, in hilly areas across the whole Arctic, where the permanently frozen soil has behaved like rock, the melting turns vast areas to mud and causes devastating and sometimes deadly landslides. The melting tundra also increases river flow, particularly in Siberia, pouring large quantities of fresh water into the sea with so far unknown consequences to the pattern of ocean currents.

Oil and gas pipelines have also suffered leaks as their permafrost foundations, once thought to be as solid as rock, have sagged. Part of the Trans-Siberian Railway has also had to be rebuilt. The melting has also caused what are called drunken forests, where the trees lean over at crazy angles as the ground has sunk unevenly.

The tops of mountains and the edges of the Arctic Circle where the ice and permafrost

Opposite: Animals have evolved and adapted to occupy habitats that are extremely inhospitable. The polar bear is the top Arctic predator, and its main diet is seals, which it catches on the sea ice. The disappearance of the ice, particularly its early melting in spring, is making it difficult for these bears to feed their young and is threatening their survival.

Above: The walrus does not get as much publicity as the polar bear, but it, too, is a specialist Arctic dweller and survivor in extreme conditions. It also faces life-threatening changes to its hunting and breeding grounds. In the background is the Greenpeace ship *Arctic Sunrise* in Chukchi Sea, where it was investigating climate change in the region.

Below: The great migration of thousands of caribou across the Canadian tundra to their calving grounds is one of the wonders of the animal kingdom. Their traditional migration routes are threatened both by oil exploration and by the changing climate.

131

occur are frequently uncomfortable places to live, at least for humans, and they are therefore sparsely populated. However, the Inuit way of life depends on ice, particularly the need to hunt animals that live on the ice, such as seals. Now the ice is often too thin to bear the weight of the hunters.

As on every part of the planet, there are specialist animal and plant species that have adapted to even the most inhospitable climate. Perhaps the best-known example of a species that has evolved to make the most of the ice in the Arctic is the polar bear. Apart from the obvious camouflage of the whiteness of its fur and its ability to hibernate underneath the snow, the polar bear's survival depends on the ice. Its main winter food is seal, and the seals' hunting grounds are the ice where they breed and fish through holes in the ice. The bears can swim long distances but not well enough to catch seals in the open water. If the ice fails to form, then the bears cannot hunt and therefore may starve to death or drown.

Scientists believe that polar bears will disappear across most of their previous range in the northern regions as a result of the shrinking Arctic ice cap and may become extinct in the wild in the next 50 years. At the opposite end of the world— the Antarctic—the disappearance of the ice is already having a dramatic effect on wildlife. A shrimplike creature called krill is central to the diet of fish, penguins, seals, and whales. In the summer there can be swarms of krill so large that the sea appears pink. Families of whales can eat millions of them without appearing to make a difference to the vast swarms.

But research shows that krill numbers have dropped dramatically since the 1970s, and scientists believe that the loss of sea ice is the explanation. Krill feed on the algae found under

Above: In the Antarctic these Adelie penguins stand framed on a massive iceberg. The ice is also where young penguins can grow large enough to have a reasonable chance of survival before taking to the dangerous waters. As the ice melts, this haven will disappear.

Opposite: Seals have their pups on the ice and use blowholes to fish in the ocean below. For the polar bears the ice is a platform to hunt seals, their vital spring food supply, without which they and their young could not survive. For the traditional Inuit lifestyle, seal hunting is also essential, both for food and clothing. But with climate change the sea ice is getting thinner and in some cases so dangerous that Inuit hunters have fallen through and drowned. In this picture taken in February 2002, seal hunters from Sanikiluaq, the Belcher Islands, Nunavut, in Canada's North use rods to probe the ice to check whether it is thick enough to bear their weight before they set off on a hunting trip. Across a vast area the sea ice now breaks up early and is too thin to bear the weight of hunters—animal or human.

"…the Arctic is not a barren land devoid of life but a rich and majestic land that has supported our resilient culture for millennia. Even though small in number and living far from the corridors of power, it appears that the wisdom of the land strikes a universal chord on a planet where many are searching for sustainability."

—Sheila Watt-Cloutier, former chairperson of the Inuit Circumpolar Conference and nominated in February 2007 for the Nobel Peace Prize for her work to end global warming in the Arctic and worldwide

the surface of the sea ice in winter. The ice acts as a shelter for the krill, which have a chance to thrive before it melts and exposes them to predators. Scientists from nine countries working in the Antarctic pooled their data on the species and concluded in 2005 that numbers had declined 80% in 30 years. This is a dramatic loss in the main food supply of species like penguins. Scientists believe this accounts for the depletion in penguin numbers because they rely heavily on a plentiful supply of krill to feed their young in the short Antarctic summer.

But krill and their predators are not the only creatures affected. The Southern Ocean is about 1˚C warmer than in the 1960s, and a lot of bottom-dwelling creatures, like mollusks, limpets, and scallops, are struggling to adapt. Limpets cannot turn over in warmer water, and scallops lose the ability to swim. These creatures cannot move farther south to find better conditions; the Antarctic land mass stands in their way.

It is elsewhere on the planet that the melting ice directly affects humanity. Kilimanjaro gets its name from the Swahili for shining mountain because of its snow cap; it shines no more. The glaciers and the snow that once crowned it are almost gone. Many scientists believe this is mostly because this part of Africa's climate has become drier, so there is less snow to replenish the glaciers, rather than simply because the peak is warmer. Either way, and probably within 10 years, the crater of this fantastic volcano will have lost its ice for the first time in 10,000 years.

All over North and South America, Europe, and Asia glaciers are melting at an ever faster rate. In the Italian Alps during the heat wave of 2003, 10% of the total mass of glaciers melted in a single summer. In January 2007 the Organization for Economic Cooperation and Development

(continued on page 139)

Left: Elephants and zebras in front of Africa's tallest mountain, Kilimanjaro. The name Kilima Njaro means "shining mountain" in Swahili, but its once permanent ice cap, which is such an attraction for tourists, is melting fast. Scientists say that the glaciers, which have been there at least 11,000 years, will be gone within 15 years, and then the occasional snow will melt not long after it falls. The rain falling on the lower slopes of the mountains is important to feed the springs and rivers that provide water for the big game and people who live on the plains. The loss of ice will bring unknown changes to the hydrology of the area.

Above and opposite: These photos document the loss of glaciers on the top of Mount Kilimanjaro. This mountain, which dominates its surroundings, is remarkable because of the climatic changes in such short distances from the dry tropical plains, through rain forests to cold grasslands, and finally to the ice fields at the 19,340-ft. (5,895-m) summit. In the last 40 years 55% of the ice in the glaciers has disappeared, and the process is speeding up dramatically, as these photos taken in 1998 and 2003 show. Since the mountain is only 200 miles from the equator, there is no summer or winter, but the photos were taken at the same season. In the past five years vast amounts of ice have disappeared. The mountaineer who took the photographs, Alastair Dobson, said of his experience: "I didn't recognize the top of the mountain in 2003—gone were the walls of ice my daughter and I had rested against in 1998, gone was the corridor between the cliffs of a glacier—all I could see was volcanic ash, boulders, and rocky outcrops. The trail leading to Uhuru looked like a beach—no snow, no ice, just a broad expanse of brown gravel."

"This is the biggest challenge our civilization has ever had consciously to face. If this goes on, we will lose ice cover on our planet. The process will cause such rapid transformation we will have enormous trouble adapting."

—Sir David King, the British government's
chief scientific adviser, London, March 2006

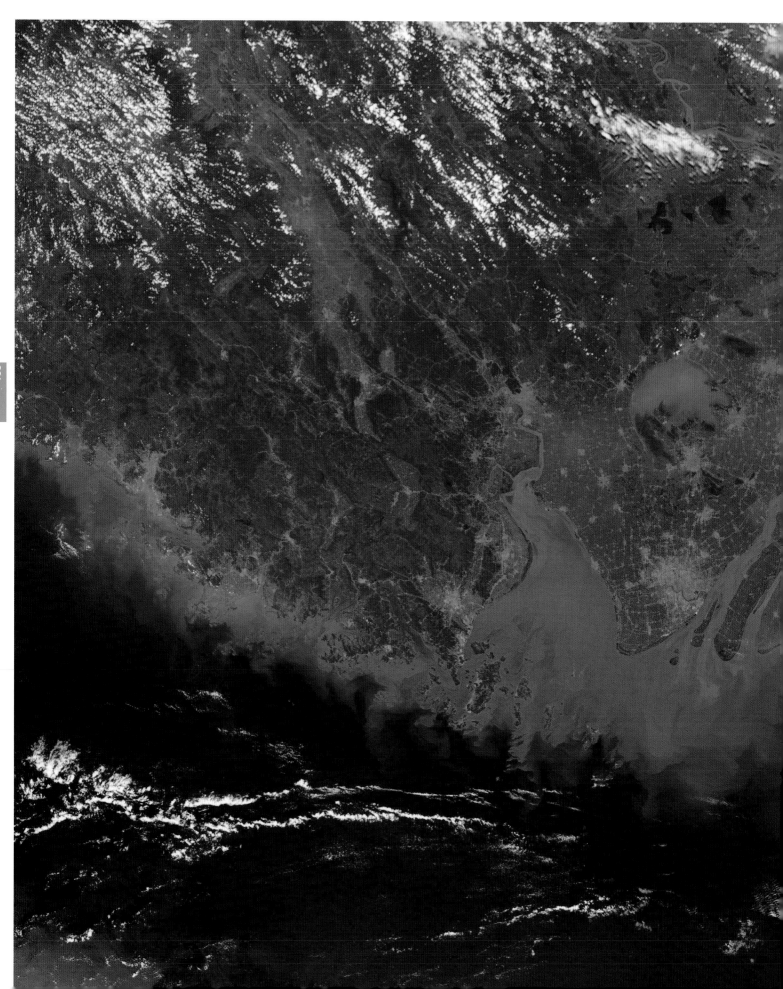

predicted that 75% of all Alpine glaciers would disappear within 45 years, closing all but a handful of the highest ski resorts.

In Europe, Africa, Asia, and the Americas the glaciers are vital for another reason—water supply. To people sometimes hundreds of miles downstream, it is not obvious that most of the major rivers in the world are kept flowing by glaciers once the spring melt of the winter's snow has finished. These glaciers are vital in arid regions for releasing water through long, dry summers.

Over the last half century, as glaciers have generally melted at an ever faster rate, the water supply in the valleys below, sometimes far away, has often seemed ever more plentiful. But this bonus cannot last; the ice is disappearing, and there are hidden dangers far up the valleys.

For example the great rivers of Asia, which provide water for the most populous regions on Earth, are fed by the meltwater of glaciers of the Himalayas. The Ganges, Indus, Brahmaputra, Salween, Mekong, Yangtze, and Huang He all owe their summer flow to the meltwater of glaciers. They provide water for drinking, irrigation, and hydroelectric schemes. As glaciers have pushed down the mountains through the centuries, they have ground out rocks and earth into giant mounds along their front edges and along their sides. As the glaciers recede, these mounds of rock and debris, called moraines, act as dams and create freshwater lakes where there was once ice. These can be incredibly dangerous because the dam walls are often partly frozen at the base, and as they melt, the whole structure collapses. In the steep Himalayas this can lead to catastrophe, as a wall of water, mud, and rocks hurtles down the mountainside. In 1985 a glacial lake burst in Khumbu, Nepal, killing at least 20 people.

The latest surveys of 3,300 glaciers in the Nepalese Himalayas show 2,300 of them have glacial lakes in the valleys where they have receded. These lakes are continuing to grow because of rising temperatures. Jennifer Morgan, director of the World Wildlife Fund's global climate change program, says: "Himalayan glaciers are among the fastest retreating glaciers globally due to the effects of global warming. This will first increase the volume of water in rivers, causing widespread flooding dangers. In a few decades this situation will change. Water levels in rivers will decline, eventually resulting in water shortages for the hundreds of millions of people who rely on glacier-dependent rivers. This threatens massive economic and environmental problems for people in western China, Nepal, India and Bangladesh."

The latest report from the Intergovernmental Panel on Climate Change (IPCC), published in April 2007, predicts increased flooding, rock avalanches from destabilized slopes, and water loss in the next 20 years, followed by decreased river flows as glaciers recede. Freshwater availability will decline, and this will affect more than 1 billion people by the 2050s.

When I visited Georgia in spring 2006, the effects were apparent. The villages on the southern side of the Caucasus had been placed on alert because of the danger of flooding and mud slides. Although part of the problem was deforestation because impoverished people had cut down the trees to try to keep warm, climate change was also to blame. The homes of 400,000 families in 3,000 villages were at risk, according to Professor Emi Tsereteli of the State University of Georgia. In the previous 35 years the glaciers of Georgia had retreated 27%. They had an important role in the summer flow of two of the country's major rivers, Inguri and Tergi.

Left: China's battle to control the flow of its rivers, in this case the Yangtze, begins in the Himalayas and ends here in the Yellow Sea, where the sediment-rich plume can be seen from space. The snowmelt and glaciers account for the spring flood and much of the flow of the river in summer. As the glaciers recede farther and farther each year, it appears far downstream that the summer flow is plentiful, but when the mountains warm sufficiently and the ice finally disappears, the water supply to millions will be put in jeopardy.

Opposite: An early example of what can be done to improve the quality of life in the developing world without damaging the climate. The Himalayan village of Namche Bazar in Nepal has had electricity since 1983 when the Austrian government paid for a hydroelectric project so that lightbulbs could replace kerosene lamps. The village is surrounded by towering peaks, which are the watershed for the rivers that bring life to China on one side and to India, Pakistan, and Bangladesh on the other.

Top: Scores of Cambodian fishing boats gather at dawn on the Mekong River for the high point of the fishing season on January 4, 2004. Huge fish migrations every January, caused by the annual flood cycle of the Mekong, provide a vital source of protein to Cambodia's 13 million people, although conservationists fear that upstream damming threatens the long-term future of the delicate ecosystem. The health of both rivers and their flow are directly affected by the climate changes in the mountains.

Bottom: Participants with life buoys and balloons swim across the Yangtze River at the opening ceremony of the 2005 Wuhan International Yangtze River Crossing Challenge in central China's Hubei Province on June 18, 2005. More than 1,100 enthusiasts from all over China took part in the 6,560-ft. (2,000-m) floating contest at the ceremony.

"Many people have falsely assumed that you have to choose between protecting the environment and protecting the economy. Nothing could be further from the truth. In California, we will do both."

—Arnold Schwarzenegger, governor of California

Professor Tsereteli said that as a result of the changes, "mud flows, landslides, flooding, and other gravitational processes are costing Georgia at least $150 million a year in damage, and it could easily reach as much as $1 billion." He said this was not a new problem for Georgia. "Over the last 30 years 50,000 families have been relocated because of environmental disasters, but every year Georgia is facing a worsening situation." One of the problems he most fears is mud slides caused by melting glaciers, which cut into the hillside. They have the biggest potential for a catastrophic accident like the one on the Caucasus in southern Russia that was caused by the collapse of the Kolka Glacier in September 2002. An avalanche of ice, snow, and rocks hurtled 8 miles and swallowed the village of Nizhniy Karmadon and other settlements, killing at least 125 people. This could just as easily happen on the Georgian side of the Caucasus, he said. He also thought that the highway that runs north through the mountains linking Georgia with Russia could be cut by landslides, as could a gas pipeline along the same route, which is vital to the region's energy supply. The pipeline had been threatened by two landslides in 2005.

At the other side of the world, in Peru, the story is the same. Using the vast Andean ice caps and glaciers, 70% of Peru's power comes from the hydroelectric dams catching runoff, but officials fear much of it could be gone within a decade. At the same time, new mountainside lakes are bulging from the melt, threatening to break their banks and devastate the towns below. The IPCC says in Latin America water availability for human consumption, agriculture, and power generation are all going to be seriously affected as rainfall is reduced and glaciers melt.

But this is a worldwide phenomenon, and there is no way that even the richest and most powerful can shield themselves from the effects of climate change on ice. California, the Sunshine State, probably has the most sophisticated large-scale water-control systems on the planet. It also has a carefully studied climate. How much it snows in the Sierra mountains each winter means a lot more to the 36 million Californians than how good the skiing will be. The spring and summer meltwater is carefully collected and shared. It is vital to industry, agriculture, domestic users, and the wildlife. California's Department of Water Resources has already predicted that demand would exceed supply by 2020—and the shortfall would be the equivalent of a year's supply to 11 million people. That was before a report by the National Academy of Sciences in August 2004, which said that climate change would cause a serious decline of the Sierra snow pack. The report said that depending on how much worldwide emissions of carbon dioxide were cut, warming would reduce the winter snow by between 30% and 90% by the end of this century. The runoff from streams would drop by at least half.

The 19 researchers from leading U.S. universities studied the snow pack because they said the volume of water produced by the melt affects 85% of the drinking and irrigation water available for Californians. In addition, snow-fed hydroelectric plants currently produce about a quarter of California's power. This knowledge about how much carbon emissions will affect local climate has caused Arnold Schwarzenegger, California's governor, to adopt an 80% greenhouse-gas reduction target, the most ambitious of any state. But in most areas of the world, this political action is missing. That is probably because the direct connection between global warming and the water supply is not being made, so the sometimes imminent threat to the future well-being of local people is still not appreciated.

> "Over the past three decades, more than a million square miles of sea ice—an area the size of Norway, Denmark, and Sweden combined—has disappeared."
>
> —Natural Resources Defense Council

Above: This is what every Californian wants to see each winter, and it is not just for the skiing resorts outside Los Angeles. This photo was taken in December of 2002 when 12 ft. (3.5 m) of snow fell in the Sierra Nevada. The annual snowfall is the lifeline for the Sunshine State's vast agricultural industry, and the spring runoff is carefully collected in dams to provide the summer water supply for farmers and the people of a state that would otherwise be parched.

Above and opposite: A series of before-and-after photos documenting the retreat of glaciers in Svalbard, Norway. This is well inside the Arctic Circle and the most northerly point washed by the warmer waters of the North Atlantic Drift. The water begins its journey in the Gulf of Mexico and brings heat to western Europe as far north as Svalbard, where the sea temperature in 2006 was a full 2°C warmer than normal. The color photos at the bottom were all taken in 2002 when a Greenpeace expedition sought to put volunteers in the landscape in the same place as the Norwegians were situated in the originals above them. All the photos are of different views of the giant Blomstrandbreen Glacier in Kongsfjorden. The earliest, on the far right, was taken in 1918, with all of the rest taken in the 1920s. Large areas of black rock covered since the last Ice Age, 11,000 years ago, are now absorbing heat from the sun, adding to global warming and accelerating the ice melt.

Rising Tides

Previous spread: The hurricane season begins in June and now lasts almost half a year in the Caribbean and along Florida's coast; 2005 was the worst storm season ever recorded. Key West resident Gregorio Nodal walks in the flooded North Roosevelt Boulevard after Hurricane Wilma hit Florida's southwest coast on October 24, 2005, two months after Hurricane Katrina devastated New Orleans. Hurricane Wilma crashed ashore in southwest Florida, again bringing flooding before roaring across the peninsula, pounding Miami, Fort Lauderdale, and West Palm Beach. It had previously hit Mexico's Yucatán Peninsula and killed 17 people across the Caribbean.

Above: The locals call them King Tides, and their children see them as fun as they play in the incoming sea as it sweeps across their islands. But many fear it is the beginning of the end for the Buota village, on Tarawa Island, one of the many vulnerable settlements on the scattered atolls of the nation of Kiribati in the Pacific Ocean.

Above: These low-lying coral islands, which currently rarely flood, are expected to be completely inundated as sea-level rise continues. These photos, taken in February 2005, show that preserving freshwater supplies and growing vegetables in uncontaminated soil are becoming increasingly difficult.

Sea-level rise is no distant threat. Already some low-lying islands are having to be evacuated and coastlines are being squeezed all over the world, a process that scientists agree will continue for hundreds of years. So the main battle is to understand how much the sea will rise and how fast. How long is it before large areas of some of the most populated parts of the Earth have to be abandoned to the sea?

As with so many other aspects of climate change, there are uncertainties about how quickly the problem will grow into a large-scale disaster. Imagine the upheaval for the countries involved. What happens to a nation's stability when the vast populations of low-lying Bangladesh, the Mekong Delta in Vietnam, and the Nile Delta in Egypt are forced to move? In each case these delta dwellers work the fertile farmlands that feed the nations concerned. Much of this farmland will be inundated. For these countries and many others, the important question seems to be how quickly the waters are rising and what they can do about it.

For the layman the most obvious cause of sea-level rise is the melting glaciers and the huge ice mountains of the Arctic and the Antarctic, now well documented across the world. The water runs downhill and meets the sea, like a dripping tap filling a bath.

Scientists, however, calculate that the main current cause of sea-level rise is the warming of the oceans. Water expands as it warms. The atmosphere is being warmed by the man-made greenhouse effect, and because of the vastness and depth of the oceans, it takes a long time for the water to absorb the heat. This is why scientists say that whatever we do now, the oceans will go on rising for 300 years and possibly 1,000 years. We can, however, take action to slow down the process by keeping the temperature rises as low as possible, with the aim of ultimately bringing them down again.

Working out how much oceans rise because of thermal expansion is complex because warm water expands much faster than cold. At 5˚C a rise of 1˚C causes an increase in volume of only one part in 10,000. However, at 25˚C, a temperature found at the surface all across the tropics, a rise of 1˚C causes a rise of three parts in 10,000. If this 1˚C rise happened across the top 328 ft. (100 m) of ocean in the tropics, it would increase the depth by 1.2 in. (3 cm).

Although logically, because water is liquid, sea-level rise should be the same everywhere; however, scientists do not believe this will happen. It depends partly on sea temperatures in different places and changes in ocean currents and air pressure, which can lead to dramatic rises and falls in sea levels. Some areas may see very small rises and others very large. So far the computer models are still unable to predict where these changes might occur. About a third of the sea-level rise in the last century is thought to have been caused by thermal expansion—and everywhere, sea temperatures are rising. The effects are bound to be increasingly significant.

The next big question is, What is happening to the two biggest lumps of ice on Earth, the Greenland and Antarctic ice caps? Some scientists at first believed that while some melting is occurring, it should be discounted because a warmer world will produce more water vapor and therefore more snow. This would make the ice caps bigger, not smaller. But the evidence is that this is not happening as much as scientists thought it would. Melting is speeding up faster than the snow can build. However, the calculations of what this will mean for sea-level rise are as difficult as for thermal expansion. Both ice caps are so enormous that they would not disappear completely for many centuries, and probably millennia. Only small fractions of them would have to melt to cause alarming sea-level changes. Even a 3-ft. (1-m) rise in oceans would radically alter coastlines across the world and turn thousands of square miles of productive farmland into new seas or salt marsh.

Opposite: The Nile River and its delta as seen from space. The Sinai peninsula and the Red Sea are on the right. It shows how little of this vast landmass is occupied by human settlements and how Egypt is so heavily reliant on the Nile Delta for growing the food that feeds the country. The construction of the Aswan Dam, completed in 1970, cut off the silt that the annual flood brought to the Nile valley, which fertilized, replenished, and extended the delta into the Mediterranean. Now sea-level rise is gradually inundating the delta, reducing Egypt's ability to grow food.

There are a lot of uncertainties, and until recently many scientists believed that the melting of Greenland and the Antarctic was not a threat, at least not for another 100 years, but similar to other areas of climate science, as time has passed, the news has continued to get more alarming.

Recent work for the Intergovernmental Panel on Climate Change (IPCC) shows that mankind has been over-optimistic in placing many of its largest settlements so close to current sea level, assuming that this will not change. During the last Ice Age the sea was 394 ft. (120 m) below the present level because so much water was trapped in the ice caps. As the ice retreated between 15,000 and 6,000 years ago, sea levels rose about 0.4 in. (10 mm) a year on average before slowing right down and remaining more or less unchanged for the last 3,000 years.

But in the 20th century sea-level rise began again. IPCC scientists were divided about how much and how quickly the oceans would warm and the ice would retreat in the face of higher global temperatures. Nearly 10 years ago they came up with a figure of between 4 in. (11 cm) and 30 in. (77 cm) in the 21st century but have since revised it upward. Information in 2005 from 177 tide-gauge stations with a global range shows that over the last 55 years sea level has increased approximately 0.6 in. (1.7 mm) a year. Over the last 10 years it speeded up again to more than 0.8 in. (2 mm) a year. But new research published in 2006 says even this is an underestimate. Eric Rignot of NASA's Jet Propulsion Laboratory in Pasadena, California, and Pannir Kanagaratnam of the University of Kansas used satellite data to monitor the speeds of glaciers that drain the Greenland ice sheet over a 10-year period up to 2005. There were large increases, first in southern Greenland, but

after the year 2000 also in the north. Several of the largest glaciers have doubled in speed over 10 years and are now advancing at 9 miles (14 km) a year. This has increased the amount of ice being dumped in the sea from 22 cu. miles (90 cu km) in 1996 to 54 cu. miles (224 cu km) in 2005. This is enough to cause an annual sea-level rise around the world of half a millimeter. The researchers estimate that global sea-level rise has now increased to 3 millimeters a year.

Separate calculations show that increased glacier and ice cap melt could increase sea levels by 6.5 ft. (2 m) this century, according to the Sir Alistair Hardy Foundation, which supports a group of ocean scientists based at Britain's Southampton University. Add to that an addition of up to 16 in. (40 cm) for expansion of the seas because of warmer water, and sea-level rise will have potentially catastrophic consequences for low-lying regions.

Although these new figures seem to be high, they are still low compared with the potential for sea-level rise. Scientists have calculated how much ice, and therefore how much water, is contained in glaciers and various ice caps. If all of the world's glaciers melted—and the vast majority are already in serious retreat—then sea level would rise half a meter. If the Greenland ice sheet melted, it would rise an additional 20–23 ft. (6–7 m), and the West Antarctica ice sheet another 26 ft. (8 m). The West Antarctica ice sheet is vulnerable because it is grounded on rock well below sea level, so there is potential for the sea to "float" it off, causing it to disintegrate. The speed with which this can happen was illustrated when the giant Larson B ice shelf in the Antarctic Peninsula disintegrated in a single summer. If all of Antarctica melted, sea-level rise would be 260 ft. (80 m)—although that is said to be so far off as to be beyond worrying about.

Left: The United Nations building in New York in 2005. The picture shows how vulnerable the city is to sea-level rise. The basements of skyscrapers on the lower parts of Manhattan are likely to be flooded before the end of the century, rendering worthless some of the most expensive real estate in the world.

What has changed in the last five years is new observations in Greenland and Antarctica. Temperatures have risen over Greenland and the Antarctic Peninsula far faster than elsewhere on the planet, sparking a much speedier melting of ice than was anticipated. Huge pieces of ice shelf have floated away from the Antarctic Peninsula, in total around 5,200 sq. miles (13,500 sq km) in 50 years, leaving giant glaciers on the mountains with unhindered passage into the sea. In Greenland the ice cap is 7,000 ft. (2,100 m) deep. It is already melting around the edges up to 3,300 ft. (1,000 m) above sea level in summer and exposing rock that has not been seen since before the last ice age, at least 20,000 years ago.

Professor Niels Reeh, from Denmark, who has been studying the Greenland ice for 20 years, has a different set of figures for this ice cap. He says ice losses in the four years between 1995 and 1999 were about 10 cu. miles (50 cu km) of ice a year. This is sufficient to raise sea levels worldwide by 0.13 millimeters a year. But he agrees that since then the rate of melting has increased as temperatures have continued to rise in the region, and they may already have reached half a millimeter a year.

What is happening in Greenland is a perfect example of what constitutes dangerous climate change. The EU maximum limit of a rise of 2˚C is a global average, but the rise in Greenland is expected to be one and a half times the average. At the same time, scientists calculate that if the temperature in Greenland were to rise by 2.7˚C, then it would trigger irreversible melting of the ice cap. In other words, a world average limit of 2˚C would still be too high to avoid a long-term but unstoppable sea-level rise of 23 ft. (7 m). In 2005 scientists meeting at the international climate change conference in Exeter revealed for

the first time that in addition to the Antarctic Peninsula and Greenland, ice shelves in West Antarctica were melting, too. This had been previously thought to be unlikely to happen for 100 years, but scientists using satellite calculations now think 60 cu. miles (250 cu km) of ice is being added to the sea each year, making a considerable contribution to the 3–millimeter annual increase. Work is still under way to understand the forces causing the problem.

Apart from the increased temperature, what has speeded up these processes? In the case of Greenland and other glaciers that grind their way across rocks, it appears that meltwater can act as a lubricant. Meltwater seeps down through a glacier to the bottom and allows the ice to slide much faster over the rock, often tripling the speed as it reaches the warmer melting zones or breaks off into the sea to form icebergs. In West Antarctica and elsewhere rising sea temperatures under ice shelves both melt and lubricate the moving ice, again dramatically speeding up the process.

While uncertainties persist, it is clear that this is another area where scientists have underestimated the speed of change, perhaps unwisely. It can lead to complacency in those whose responsibility it is to take action to combat the problem. As with global averages in temperatures, tiny increases in sea-level rise each year do not seem to pose much threat. But while in a calm sea a 2–millimeter level change cannot even be noticed, it is different when magnified by storms, wind, and tides. A combination of wind and tide is responsible for the serious flooding in Venice and St. Petersburg. In eastern Britain and the Netherlands it is a northerly gale pushing the water down the North Sea to the bottleneck of the English Channel that is the danger. If this coincides with a high tide, a massive warning

(continued on page 158)

Right: People on wooden walkways in St. Mark's Square flooded by high water in Venice, November 14, 2001. At the beginning of the last century, flooding occurred in the square, the lowest part of Venice, once or twice a year. Now the water comes over the banks of the Grand Canal and up through the paving stones and the mosaic floors of St. Mark's cathedral more than 100 times a year. Venetians are abandoning the city as living conditions at street level become increasingly hazardous.

Above and opposite: Venice may be the most famous example of a city facing inundation, but many of the world's most populous and richest cities are on low-lying coastal plains and are vulnerable to sea-level rise. Although most coastlines have remained stable over the last 10,000 years, that is only because there has been no change in sea level. Without new sea defenses to block the rising tides, all of these cities face waterfront flooding. London and Rotterdam already have barriers against the sea but need to build new ones to deal with tidal surges in the North Sea that can raise sea levels by as much as 16 ft. (5 m). St. Petersburg is also building its own barrier. Apart from the loss of life that a storm surge and flooding would bring to any of these cities, the economic consequences would also be vast. The world's major financial institutions and most expensive real estate, housing thousands of the world's highest-paid workers, would be wrecked.

St. Petersburg

Hong Kong

Bombay
(Mumbai)

Singapore

network is activated. The surface of the sea can rise 16 ft. (5 m) above normal, potentially over-topping the defenses that protect millions of people. The same dome of water occurs under the intense low pressure of a hurricane. Add to that the height of storm-driven waves and a high tide, and the extra millimeters are multiplied many times into a more serious threat. The story of Hurricane Katrina and low-lying New Orleans could be repeated many times. Much of the southern United States, particularly Florida, is vulnerable to such storms and will increasingly be so.

But flooding aboveground is not the only threat. Many countries use boreholes for fresh water. When too much fresh water is taken out—that is greater amounts drawn off than are replaced by rain—the levels of groundwater drop. On the coast this can have two disastrous effects. One is to cause the ground to sink, and the second is to suck in salty water from under the sea. The result is both a loss of the water supply and a greater danger of flooding. Many places in the world, including parts of China, have sunk by as much as 6.5–10 ft. (2–3 m), so much so they are now below sea level and have to be defended with giant banks.

The southeast of England has another unrelated factor that adds to the danger. During the last Ice Age the British Isles were only partially covered by thick ice. As a result, the northern half of the landmass was pushed down into the Earth's crust. Once the ice melted, the North began to rise again, tilting the South and East downward. The process still continues, so to add to the danger of sea-level rise, another 1 millimeter a year of sinking needs to be factored into the calculations of London's flood defenses.

So it is no longer a question of whether parts of New York, Hong Kong, or Shanghai and dozens of other port cities will go under but when, unless they can construct robust sea defenses. A 3-ft. (1-m) sea rise would put Shanghai, the great boom port city of China, one-third under water, drowning the bottoms of the skyscrapers in the vast commercial center. Hong Kong would face a similar fate. Even London, with the best sea defenses in the world, is having to buy time by building yet more sophisticated and higher barriers to keep out the North Sea tides. Rotterdam, the biggest port in Europe, also has gates it can close when the sea threatens. Venice, on the Adriatic, which is already flooded more than 100 days a year and St. Petersburg, on the Gulf of Finland, which is sited on a former marsh, are building their own barriers to protect themselves.

The reason most of these cities grew up where they are is exactly their proximity to the sea. It is no surprise, therefore, that both the most valu-able and vulnerable parts of these commercial centers are closest to the waterfront. Their populations and businesses will have to be relo-cated inland. How soon this will begin to happen is one of the unanswered questions. At the moment, where they are able—for example, on relatively narrow river estuaries—city authorities are putting their faith in building sea defenses. For weeks and months at a time, they can continue without any threat, but at seasonal high tides or during storm surges they must have protection. This usually takes the form of huge gates designed to hold back the waters when floods are forecast. But some places, like New York, are too exposed to defend, and it is in these locales where local governments are realizing that their best hope is to slow down sea-level rise as

(continued on page 162)

Right: The Thames Barrier, London's bastion against a storm surge in the North Sea, which would overwhelm billions of pounds' worth of property and flood parts of the underground railway system. Sea-level rise means that even this sophisticated barrier is not going to be enough to save London, and already the UK's Environment Agency is working on massive new defenses. If nothing is done, the barrier would have to be lowered on every tide by the end of this century to prevent daily flooding.

"Those who will suffer the most are the small island states, who neither have the space to relocate people and the infrastructure nor money to invest in protective measures...the challenges are daunting and time is running out. We are fast approaching the point of no return."

—Maumoon Abdul Gayoom, president of the Maldives, 1998

much as possible. This means reducing their emissions and campaigning for everyone else to do the same. This, they hope, will slow temperature rises and thus keep sea-level rise in check for as long as possible, although as we have already mentioned, some sea-level rise is inevitable because of emissions over past decades.

But cities, with their large populations and expensive real estate, have less of a problem than many low-lying island states. Many island states have no high land to retreat to and face partial or complete extinction. Ultimately, entire populations will have to move to new countries and face the loss of their culture and national identity. The best-known example of these countries to the developed world is probably the Maldives. To many they are the idyllic palm-fringed vacation islands in the Indian Ocean, but to the people who live there, the Maldivians, they are a nation under siege from the sea.

The president, Maumoon Abdul Gayoom, has been a tireless campaigner on climate change since April 1987 when a storm caused severe damage to the islands. In October that year, in a speech to the United Nations in New York, he was the first head of state to alert the world to the danger that sea-level rise posed to coral island states like his own. He has since made countless speeches and written a book, *The Maldives: A Nation in Peril,* and distributed it as widely as possible. Realizing that the rise of seas is inevitable, although he still hopes it can be reversed in time, the country has a "safe island" policy, which will mean greater investment in sea walls, more solidly constructed buildings, elevated areas for vertical evacuation, and environment protection zones. The capital, Male, has a sea wall, built with Japanese aid costing $63 million. John Bennett, from New York, a representative of the United Nations Environment Program, who

spent six months helping the islands recover from the 2004 tsunami, said: "From the highest levels of Maldives society to the average person in the street, there seems to be an unusual degree of awareness about the threat that climate change poses to the country. For example, one night a taxi driver asked me if it was true that the entire country would be under water in fifteen years.

"The government has been developing and promoting a plan to build safer islands. But the destructive pattern of the tsunami waves showed that determining the characteristics of a safer island is not so simple. And persuading residents to leave their home islands, however attractive the offer, may prove challenging. Still, it is important for the government to be taking the threat of sea-level rise seriously and exploring options.

"The Maldivians are very resourceful, but unless we in the rest of the world change course dramatically, the country stands to pay a horrible price for our sins. How can we care, as we have, for the tsunami victims of today and yet turn an indifferent eye to the trouble that will befall their children and grandchildren because of sea-level rise?"

But the people of the Maldives are not alone. They are one of 43 tiny island states from the Caribbean to the Pacific via the Mediterranean and the South China Sea that are among the most vulnerable to climate change. The first of these to go will be those like the Maldives, which are composed of many low-lying coral islands and are already vulnerable. Although the tsunami disaster that struck the Maldives at Christmas 2004 was caused by an earthquake, the damage caused by tidal waves was not dissimilar to that expected from tropical storms. Waves between 3–16 ft. (1–5 m) high reached

Previous spread: The first glimpse of the Maldives most visitors get is of this exquisite turquoise-ringed spread of atolls set against the deep blue of the Indian Ocean. Depending on your point of view, this picture could show you a group of islands and a nation extremely vulnerable to sea-level rise or an idyllic paradise vacation destination.

Top and bottom left: The Maldives was known to travelers even in ancient times. Today many of the more developed islands, like Reethi Rah (top), offer five-star luxury and the vacation of a lifetime to those who can afford it.

Bottom right: The capital of the Maldives, Male, is home to a nation with a rich heritage dating back over two millennia. Today the government is battling with the problem of sea-level rise by fortifying a number of islands against the sea as safe havens when storm surges threaten.

the area, and some islands were completely washed over. According to the United Nations Environment Program, 69 of the country's 199 low-lying inhabited islands were damaged, 53 severely. Twenty were largely devastated and 14 had to be evacuated altogether. A third of the country's population, about 290,000 people, suffered losses and damage. These figures give some idea of what a combination of sea-level rise and tropical storms could do to the islands.

President Gayoom refused to talk to me about the possibility of evacuating his country and is putting all of his faith in adaptation.

"This is a scenario we do not have the means to prepare for, although as a country in the front line of climate change impacts, we accept we would be the first hit and probably the worst hit," he said. "However, we may not be the only victims as time goes on. I believe the onus is on the entire international community to deliberate on how to deal with environmental refugees."

As President Gayoom said, his country will not be alone. The Marshall Islands, stretching almost 1,000 miles (1,609 km) across the Pacific, like the Maldives, are likewise nowhere more than 6.5 ft. (2 m) above sea level and mostly less than half that. Among the nations threatened just in the Pacific are the Cook Islands, Fiji, Kiribati, Nauru, Papua New Guinea, the Solomon Islands, Tonga, Tuvalu, Vanuatu, and Western Samoa. All will lose valuable land, and some will disappear altogether. Already in Tuvalu some of the causeways connecting islands have been inundated. Families in the worst affected islets have been given sanctuary in New Zealand.

Some other islands are already untenable. In October 2005 the decision was made to evacuate the Carteret Islands, which are also in the South Pacific. For 20 years the 2,000 islanders have fought against the ocean, building sea walls and trying to plant mangroves to hold back the sea and build up the soil. Each year the waves surged in, destroying vegetable gardens, washing away homes, and poisoning freshwater supplies. The only food plants they were able to grow were planted in tubs above ground level. The islands are part of Papua New Guinea, and the government finally decided to move the islanders to the larger, nearby Bougainville island, a four-hour boat ride to the southwest.

In December 2005, at the climate talks in Montreal, it was announced that another small community living in the Pacific island chain of Vanuatu had been formally moved out of harm's way as a result of climate change. The villagers have been relocated higher into the interior of Tegua, one of the chain's northernmost provinces, after their coastal homes were repeatedly swamped by storm surges and waves. But in both these cases the islanders were at least lucky that they belonged to countries that internally still had somewhere to which they could relocate.

When the time comes for other countries, they will have to move everyone to the territory of another nation state. In a world of diminishing land area, this means finding another country prepared to give up its own resources and provide an entire community somewhere to relocate. It is a problem the world has not yet faced.

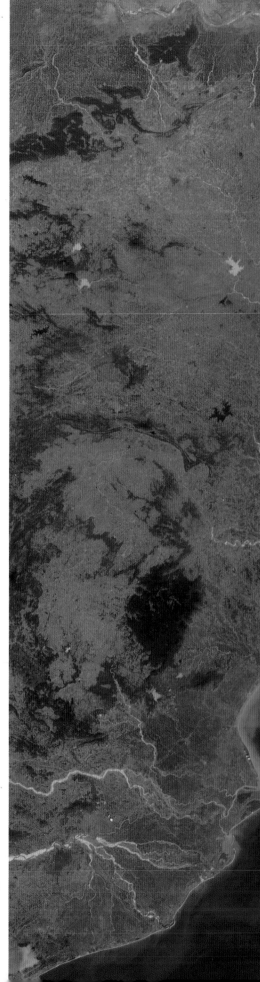

Right: When the rivers of the Bangladesh delta are in flood following the monsoons, they bring thousands of tons of silt with them from as far away as the Himalayas, adding offshore islands to this crowded country. Traditionally the poor occupy these new islands, build homes, and farm them, risking inundation in subsequent cyclones. As sea-level rise continues, this natural expansion of Bangladesh will stop and be reversed as a large part of the delta disappears under the waters of the Bay of Bengal. This picture shows 240 sq. miles (600 sq km) of the area most at risk. Along these shores and similar low-lying areas of India, more than 100 million people will be displaced. No thought has been given to what may happen to these people and the tensions it may cause in what is already one of the most crowded parts of the world.

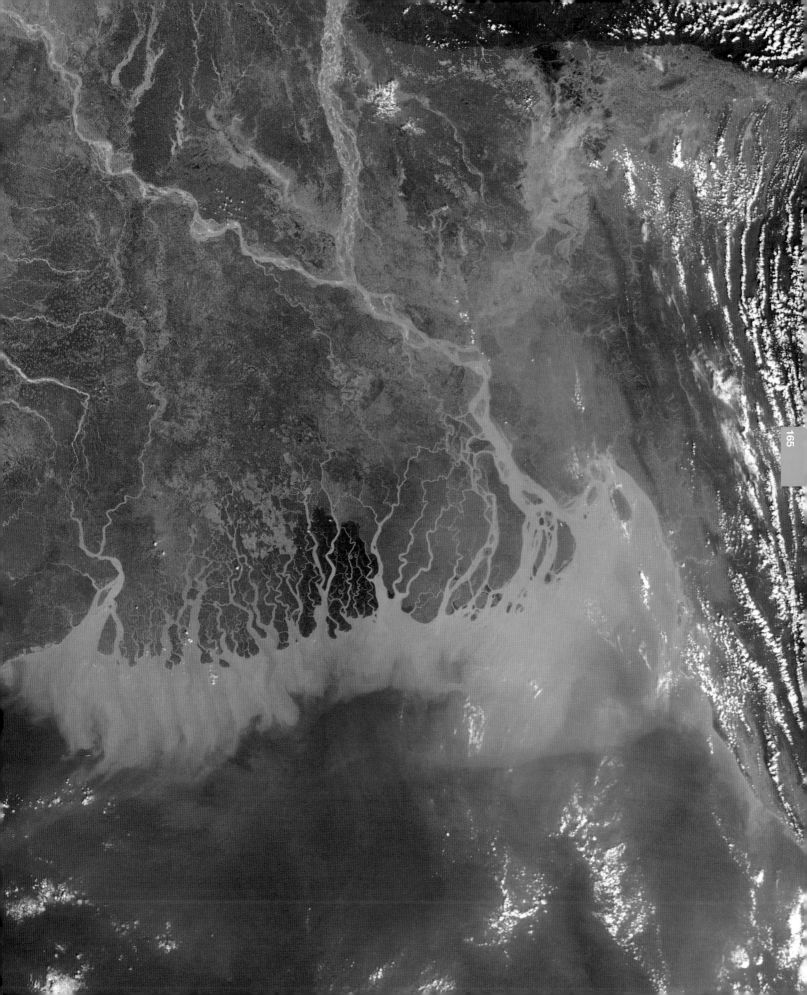

Sudden Cold?

With most debate about global warming centered on how quickly the Earth is heating up, it seems odd, not to say impossible, that some parts of the world might suddenly get much colder. But scientists have always said that climate change could involve some nasty surprises, and one of them could be western Europe being plunged into extreme cold.

Analysis of ocean sediments shows that several times in the geological record, average temperatures have dropped as much as 10˚C in as little as 10 years in Greenland as a result of dramatic changes in the ocean currents.

These currents normally distribute heat around the world. The North Atlantic is kept at least 5˚C warmer than it would otherwise be because of the circulation of warm water from the tropics, which washes far up into the Arctic Circle. Places as far north as Siberia and Alaska, which would be icebound, have far milder winters as a result.

These ocean currents act like a conveyor belt. On the surface water moves north at about 4 m.p.h. (6 kph) transporting water, warmed to 27˚C by the tropical sun, thousands of miles up the coast of Spain, France, Britain, and Norway. It is like having a giant lukewarm radiator kept on all year round, gradually distributing its heat as it moves north. When the current reaches the Arctic, the air cools the water sufficiently for it to sink and for ice to form on the surface. When the ice forms, the salt is left and makes the water underneath the ice denser, and thus heavier, than the fresh water around it. This cooler, saltier, and denser water sinks to the ocean floor, dragging in more warm water behind it. The current, equivalent in flow to 100 times that of the Amazon, is propelled back south, exactly like a conveyor belt, deep in the ocean, not resurfacing until it has traveled thousands of miles to the Southern Hemisphere. When it does resurface, it starts to warm again to begin its journey back.

The strength and speed of the current is dependent on the sinking of the heavier saltwater when it reaches the Arctic. If this did not happen, the current, popularly known as the Gulf Stream, would be turned off. In fact, the whole thermohaline circulation, as it is properly called, would cease to flow. This would have many dramatic effects on sea life because when it sinks, the current carries vast reserves of oxygen to the ocean depths and keeps many strange life forms alive. For mankind the most obvious result would be the whole North Atlantic and the countries around it being plunged into cold. It would not just make life uncomfortable; many crops would not survive because of the cold, and places like Iceland would probably become virtually uninhabitable.

This prospect has excited journalists in Europe and blockbuster filmmakers in Hollywood, and it has alarmed some politicians sufficiently to provide funds to measure whether the Gulf Stream is about to stop.

So far the results are inconclusive. The current best guess from scientists from the climate research groups at the University of Illinois and other American universities is that there is a 50% chance of a total shutdown this century if emissions are not cut. The science is complex because there are a number of contributing factors to the loss of the current, including the warming of surface waters, making them less dense; the loss of the ice in the Arctic, which would dilute the salt in the surface water and prevent it from sinking; an increase of fresh water into the Arctic from melting permafrost; more rainfall in the region; melting glaciers; increased river flow; and the loss of part of the Greenland ice cap, which would also dilute the salt. Some scientists believe that measuring devices across the Atlantic show that the Gulf Stream is already slowing down. Other computer models predict not a total shutdown but rather a reduction in the Gulf Stream of up to 50% by the end of this century. At the time of publication, however, the waters around Iceland and northern Norway are much warmer than normal. No one is sure why this is so, and it has happened in the past, only to cool down again. The warming may be due to increased hurricane activity in the southern North Atlantic, as discussed elsewhere, but scientists are still uncertain that they have understood all the factors involved in driving the ocean circulation.

Yet many researchers remain convinced that a sudden halt is possible, pointing to the fact that it has happened not once but several times in the past, although this was for natural reasons that are unlikely to occur again. The last time was at the end of the last Ice Age. When temperatures should have been rising in Europe, they plunged 9˚C and were cut all across the North Atlantic. If this happened in this century, it would be economically and socially disastrous for western Europe and eastern North America.

**Shellfish on
Death Row**

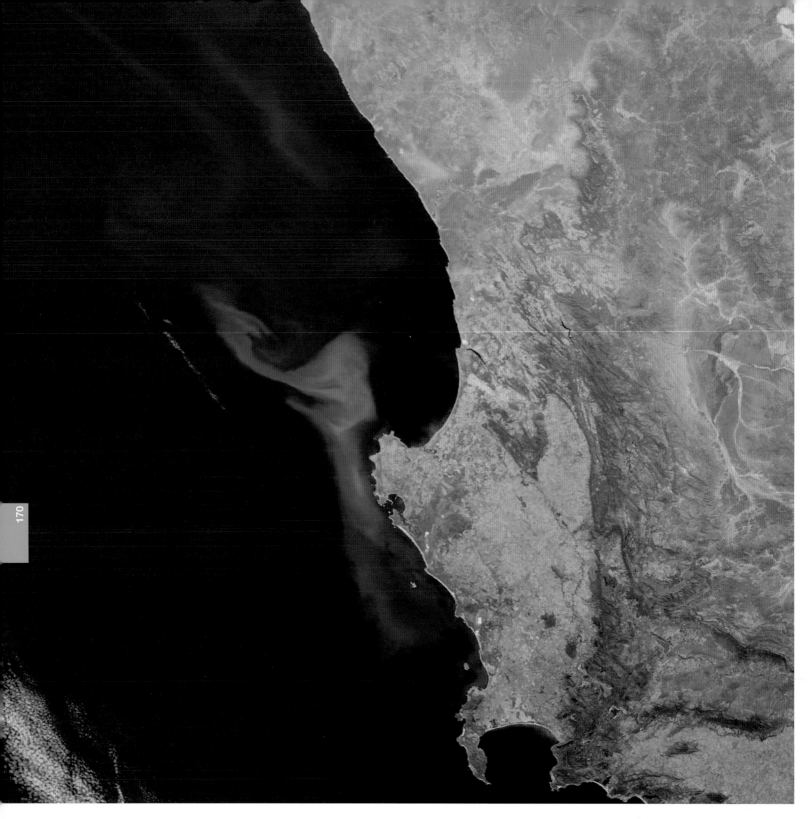

Previous spread: The oceans are under attack from climate change. One of the first major threats identified was coral bleaching. When the water gets too warm, the tiny organisms that form the coral die, leaving skeletons behind. It looks all right to the untrained eye, but parts of the Great Barrier Reef of Australia, shown here, which is normally swarming with life, are becoming ghost reefs of dead coral.

Above: A vivid turquoise streak colors the waters of the South Atlantic Ocean around Cape Columbine, South Africa. This sort of ocean coloring occurs when millions of the microscopic marine plants, called phytoplankton, grow near the surface of the water. Such blooms are common in this region, where ocean currents sweep cold water from the ocean floor to the surface. The rising water carries nutrients from the seafloor, which nourish the tiny plants. The plants in turn feed fish and other marine life. As a consequence, regions that boast frequent phytoplankton blooms also tend to support a thriving ecosystem. Many tiny creatures, which form the start of the food chain, need the ocean water to be slightly alkaline so they can extract minerals to build their skeletons. Increasing acidity in the sea caused by absorbing extra carbon dioxide threatens their survival.

171

Above: Large phytoplankton blooms are generally good news for all of the animals higher up the food chain that eat them, but too much can also create dead zones in the ocean where no life can survive. As the tiny plants die and sink to the ocean floor, bacteria begin to break them down. When the plant matter is dense, bacteria can consume all of the oxygen in the water, leaving oxygen-free areas that cannot support life. This region off the east coast of Argentina is one of three known naturally occurring dead zones in the world. The other two are off the coasts of Peru and Oregon. Other dead zones, all man-made, occur near the mouths of rivers where agricultural runoff feeds phytoplankton. The largest in the world is where the Mississippi discharges into the Gulf of Mexico.

Scientists gave warning in 2005 of a newly discovered threat to mankind. They say it could wipe out coral and vast numbers of tiny ocean creatures on which many species of fish, whales, and other sea life depend for survival.

Although the phenomenon is caused by excess carbon dioxide in the atmosphere, it is not a global warming problem, but a simple chemical reaction between air and sea. Carbon dioxide mixed with water produces carbonic acid, which is making the alkaline oceans more acidic. But for hundreds of thousands and probably millions of years, plankton, shellfish, and corals have adapted to use the stable levels of calcium and carbon in the sea to make their shells.

This acid threatens to alter the balance of marine life. Chalk, perhaps most famously seen at the white cliffs of Dover, is made from the remains of trillions of these creatures, which still form a vital part of the present food chain.

So alarmed did marine scientists become about this new threat caused by the extra carbon dioxide in the atmosphere that in 2005 special briefings were held for British government departments. Carol Turley, head of science at Plymouth Marine Laboratory, warned them of a "potentially gigantic" problem for the world. "It is very urgent to warn people what is happening," she said. "Many of the species we rely on to eat, like cod, will disappear. The whole composition of life in the oceans will change."

Later research showed that while all of the world's oceans would eventually be affected, there were certain areas that were more vulnerable. These were the Southern Ocean around Antarctica and the tropics where corals form. This is mainly due to the way that the ocean currents distribute the water with greater carbonic acid. In the shallow seas where corals form, the impact of carbon dioxide is likely to be greater, and the tiny organisms that create the reefs may no longer be able to do so because not enough will survive.

On this page: These are the vibrant colors and teeming life associated with healthy coral reefs, the rain forests of the oceans where there are thousands of different species of animals and plants living together in a complex ecosystem. At the top a type of Gorgonian coral with its colony of fish in a still healthy part of the Great Barrier Reef in Australia. At the bottom and opposite are scenes from the Philippine coral community. These creatures include a bulb-tentacle sea anemone and a shoal of fish known as monos.

"Some scientists are saying that, in 35 years, all the coral reefs in the world could be dead—it could be less or more. Put it this way, my children may never get the opportunity to go snorkelling on a living reef. Certainly, my grandchildren won't."

—Jerry Blackford, joint author of a paper on the acidity of the oceans, presented to the Exeter conference on climate change February 2, 2005

On this spread: Trying to find more signs of life, Dr. Peter Harrison from Southern Cross University, New South Wales, is standing surrounded by bleached coral on the Great Barrier Reef, still one of Australia's greatest tourist attractions and the largest coral reef on Earth. Warmer sea temperatures are killing the reef, wiping out the organisms that build the coral. Reefs can recover with time, but not if there are repeated incidences of the water being too warm, as has happened in the last decade. Reefs are vital as fish breeding grounds and as a barrier to protect islands from storm surges.

Liquid Acidity Levels (pH)

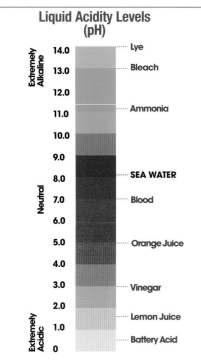

Extremely Alkaline	14.0	Lye
	13.0	Bleach
	12.0	
	11.0	Ammonia
	10.0	
	9.0	
Neutral	8.0	**SEA WATER**
	7.0	Blood
	6.0	
	5.0	Orange Juice
	4.0	
	3.0	Vinegar
	2.0	Lemon Juice
Extremely Acidic	1.0	Battery Acid
	0	

The sea has been slightly alkaline for millions of years because plants take carbon dioxide out of the water to grow. This process has kept the balance right for many shelled creatures to evolve. They take the chalky minerals out of the water to make their protective shells. In slightly acid water these shells would dissolve. The excess carbon dioxide in the atmosphere has meant the formation of carbonic acid, altering the balance between acid and alkaline. An approximate pH value of 7.0 is regarded as neutral—above that level, around 8.2, the water is alkaline and good for shellfish and coral formation. As it starts to drop toward 7.0, life gets more and more difficult for animals that rely on shells for survival.

The oceans' vital role in limiting CO_2 levels in the air will also have to be reassessed in light of these findings. Plankton are as important as plants and trees in the take-up of carbon. Scientists estimate that about half the 800 billion tons of CO_2 put into the atmosphere by mankind since the start of the Industrial Revolution has been soaked up by the sea. Much of the carbon is fixed in the shells of creatures called coccolithophores, whose bodies have been making chalk layers for millions of years. Trillions live on the ocean surface, and when they die, their shells sink to the bottom, thus taking the carbon with them. They could not survive in a more acidic sea, and their removal of carbon from the atmosphere would stop.

"These creatures are part of our survival bubble. The oceans give us a sustainable atmosphere by taking out the carbon dioxide. They're the lungs of the planet. People have not woken up to the potential impact their removal will cause," said Dr. Turley.

The acidity of liquid is measured on the pH scale, from 0 to 14, with 7.0 being neutral; the higher the number, the more alkaline. The oceans have previously recorded an 8.2 pH reading, but this has now dropped to 8.1 and is continuing to fall.

Experiments show that even a small increase in acidity reduces the ability of shellfish and plankton to grow and causes a population fall. The loss of corals would seriously affect small Pacific islands, which owe their existence to the building abilities of these tiny creatures. Reefs also play an important part in the protection of coasts. The biggest fundamental problem is the effect on the food chain. Zooplankton, essential food for fish, could suffer increasing mortality rates; starfish, whelks, and other shellfish

Above: In the Red Sea the coral reefs teem with fish and other creatures. This sea star would probably not be affected by slightly more acid oceans, but the coral on which it lives would be unable to thrive. Without its favored habitat the sea star would soon be on the endangered list, too.

Below: One of the potential winners from climate change. The jellyfish does not rely on the slightly alkaline seawater to build its body, so it would be unaffected by acid oceans. As other species die out, there is a possibility that jellyfish will multiply hugely and take over the oceans.

"Those who recognize that the fortune and fate of humankind are inextricably tied to the state of the sea understand that trouble for the oceans means trouble for us."

—Sylvia Earle, *The Final Frontier in the Blue*, Dakini Books, 1999

Above: Humpback whales are not the largest of these giant mammals but are probably the best known. Their recorded songs or voices have been released on records, and they are frequently filmed. Because of their spectacular antics and mating rituals on the sea surface, they are popular with whale watchers. They eat across the world's oceans by sieving their food through giant plates in their mouths, which act in the same way as fishermen's nets. Many of these creatures that form the whales' main diet will disappear as the oceans become more acid, in turn threatening the whales' long-term survival.

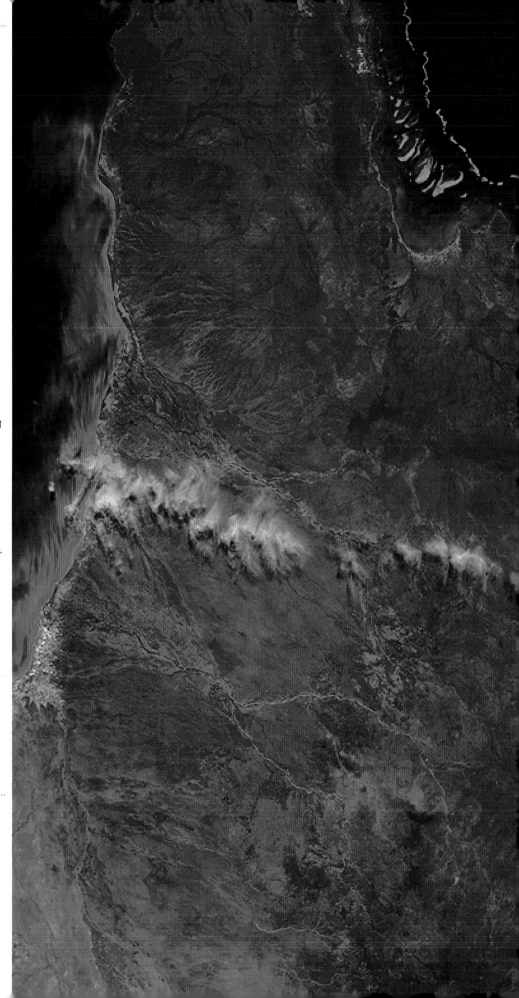

eaten by cod might perish. This could lead to population explosions of other creatures, such as jellyfish or crabs, shrimp, and lobsters, which rely on chitin rather than calcium for their shells.

As has already been discussed in previous chapters, seas around the world are warming. This has obvious effects on sea ice, and at the North and South poles on the species that have adapted to living on or underneath it. Not so well researched or understood are the changes in species that occur elsewhere. To the non-scientist it is surprising that a small change in water temperature makes a large difference to the species mix. Fishermen, of course, have long known this, simply because of the species they catch, but studying the changes scientifically is a great challenge.

Part of the reason for this is that man's other activities, apart from warming the oceans, have changed fish distribution already. Many of the world's favorite fish, like cod and salmon, are already overfished, so their disappearance from some areas has nothing to do with climate change.

However, in the North Sea, which has heavily studied shared fisheries, it is clear that climate change is making a difference to stocks. Tiny cod are dependent on certain types of plankton for food during the vital early stages of their development when they live close to the sur-face. But the warming of the surface water has meant the replacement of their preferred food with less nutritious southern varieties. The result has been less breeding success for cod, and damage to a commercially important fishery. To thrive, cod must move north.

Right: The magnificence of the Great Barrier Reef from the air disguises the fact that in many places the reef is dying. Partly to protect its fossil-fuel industry, Australia refused to sign the Kyoto Protocol and supported the United States in its attempts to destroy the only legally binding agreement to reduce greenhouse gases. If Australia's efforts succeed, the country's other great industry, tourism, will suffer. Climate change will destroy the reef and ravage the country's other natural wonders, including its forests. Droughts and fires will also be common.

Fortunately for fishermen, this has not meant empty seas. Sardines and anchovies are moving into the southern North Sea, and red mullet and bass are extending their ranges northward and westward, as far north as Norway. In the south, Mediterranean and west African species are being found off Portugal. On the opposite side of the Atlantic, off the Canadian coast, some fish species have moved south. This appears to be in response to cooler and less salty waters caused by increasing melting of ice and glaciers.

One other related problem affecting oceans is so-called dead zones. These are usually where rivers or other pollution-heavy discharges go into the sea. Perhaps the best known, and largest, is in the Gulf of Mexico where the Mississippi River reaches the sea. The main cause is excess nitrogen runoff from farm fertilizers, sewage, and industrial pollutants. The nitrogen triggers blooms of microscopic algae known as phytoplankton. As the algae die and rot, they consume oxygen, suffocating every-thing, including clams, lobsters, oysters, and fish.

In 2004 a new report by the United Nations Environment Program said that in the previous decade the number of dead zones had doubled to 150. They affected 27,000 sq. miles of ocean, an area the size of Ireland. As well as the Gulf of Mexico and Chesapeake Bay in the United States, they were also spreading to the Baltic Sea, Black Sea, Adriatic Sea, Gulf of Thailand, and Yellow Sea. They are also appearing off South America, Japan, Australia, and New Zealand and were directly linked to human settlements, agricultural development, and subsequent pollution.

Although it is a difficult problem, it is soluble. If we reduce the use of nitrogen fertilizers and prevent sewage from being directly released into rivers and the sea, natural processes can cope with the nutrient load and prevent oxygen starvation. The dead-zone problem could be made worse by global warming, which could cause changed rainfall patterns and increased temperatures. For example, U.N. scientists calculated that increased rain could cause a 20% extra flow into the sea from the Mississippi. Along with a 4°C rise in temperatures, this would cut oxygen levels in the sea by between 30% and 60%. This would massively increase the area of the dead zone.

Left: The white cliffs of Dover, one of the most famous views in Britain, are visible across the English Channel from France on a clear day. The cliffs are made entirely of chalk, which are in turn made up of the skeletons of billions of tiny sea creatures that lived in the oceans millions of years ago. Their descendants still swim in the Channel and inhabit the world's oceans. As they die and fall to the ocean floor, they create new layers of chalk. But these creatures will disappear with increasing ocean acidity because they will be unable to extract the minerals from the seawater to build their skeletons. At the same time, as with acid rain's action on many of man's ancient buildings, the cliffs will be further eaten away and undermined by the sea as high tides wash against the base of the cliffs.

Extreme Events

Previous spread: Each tropical storm gets a name—and most are quickly forgotten—but the memory of Hurricane Katrina will live for a long time, especially for the residents of New Orleans and along hundreds of miles of coastline along the Gulf of Mexico. As seen from space on August 28, 2005, before it struck the city, Katrina's swirling motion is producing winds of 160 m.p.h. (257 kph). The air pressure, another indicator of hurricane strength, is 902 millibars, making it the fourth lowest pressure ever recorded in an Atlantic storm. These two factors made Katrina a category 5 hurricane, the most violent kind. By the time the storm reached New Orleans, it had decreased to a category 3 hurricane. Even so, the water surge was sufficient to overwhelm the levees protecting the city. Because the suburbs are below sea level, the water subsequently had to be pumped out.

Above: Two men paddle in high water two days after Hurricane Katrina devastated New Orleans and flooded large areas to a depth of 12 ft. (3.5 m). Many who stayed behind drowned when levees protecting the city gave way. Thousands were left homeless in the suburbs and beyond—across Louisiana, Mississippi, Alabama, and Florida.

Above: Scenes that shocked Americans and the rest of the world for days were those of the poor who were abandoned in the flooded city without food or the rule of law. It was four days before rescuers moved in to relieve thousands of desperate evacuees and bring them out of the city, shutting down two huge shelters that had become the scene of fear and chaos. Here, in comparative luxury, a young Hurricane Katrina survivor drinks from a bottle, while others rest on camp beds on the floor of the Astrodome in Houston, Texas.

Hurricane Katrina, which wrecked New Orleans in 2005, focused the world's attention on the way extreme weather events can overwhelm man's defenses. Although the city had been warned, and was apparently prepared, it was overtaken by catastrophe. The populated areas below sea level were found to have inadequate defenses against such a storm.

Katrina was just one of a series of giant storms during the hurricane season that battered the coastlines of Central America and the southern United States. Records were broken, with more storms over a longer period than ever before. These storms became more potent because of the energy gained from the extra warmth in the oceans, particularly in the waters of the Gulf of Mexico. There were a total of 27 named tropical storms during the season, breaking the 1933 record of 21. Of these 27, 14 developed into hurricanes, again breaking the record for the most hurricanes in a season.

One of the consequences of these storms was to stimulate a long-delayed debate in the American press and on radio. These news outlets had previously been dominated by those businesses and journalists who believed that talk of global warming was nonsense. For the first time, scientists and environmental groups who believed that climate change was a threat were given a sympathetic hearing. This did not prevent the White House, however, from intervening to try to silence the scientists. Jim Hansen, the director of NASA's Goddard Institute for Space Studies, complained in January 2006 that the Bush administration had tried to censor him.

Until 2005 there had been considerable doubt in the scientific community that climate change was adding to the destructive force of extreme weather events. Problems arose because floods, droughts, and windstorms had been causing havoc repeatedly throughout the centuries. They were recorded from the earliest times, whenever history was written down, but there were no precise measurements, and the accuracy of the accounts obviously could not be relied on. The computer models of the future, developed over the last 20 years, also contain estimates

of actual conditions. They have predicted a substantial increase in intense hurricanes but with a lower level of certainty than many of the other expected results of climate change. This doubt remains in the scientific community despite the simple fact that more heat in the atmosphere equals more energy and therefore potentially more violent weather events.

To have a hurricane, or what is known as a typhoon in the South China Sea or a tropical cyclone in the Indian Ocean, the temperature of the water has to be approximately 81°F (27°C) or more. That is why, as the oceans warm, there is a potential for the hurricane season to be longer and to extend the possible range of such storms north and south. Because of this, one of the predictions in the climate models was that there was a potential for hurricanes in the South Atlantic for the first time. When the first recorded hurricane in the South Atlantic occurred in 2004, it caused a great stir in the scientific community but little comment elsewhere. This was primarily because it hit a sparsely populated part of the coast, and there were no casualties. But once again reality was catching up with the computer predictions. However, there are still some scientists who remain reluctant to say that man-made climate change has already led to more hurricanes.

The scientific reluctance to predict more dangerous storms, both north and south of the equator, is in contrast to the evidence from the worldwide insurance industry. Insurance companies repeatedly have produced figures showing that their bills for weather-related insurance claims, and not just from hurricanes, have been rising dramatically decade by decade. However, much of this is due to the increased value of assets along the coast.

(continued on page 190)

"Katrina is the first sip, the first taste, of a bitter cup that will be proffered to us over and over again. It is up to us to tackle climate change, and it does involve accepting that there is a legitimate role for government."

—Al Gore, Former Vice President, October 10, 2005

On this page: Four scenes in the days following the terrible storm that swept through New Orleans at the end of August 2005. Although the strength of the storm was predicted and the city was said to be well prepared, more than 1,000 died, and there was no help for the thousands of poor people and some tourists left behind. At top left a victim is buried using bricks and sand in a shallow grave six days after the hurricane struck. The woman, named Vera, had lain exposed to the elements. At bottom left bodies were left uncollected on the streets while survivors, with whatever belongings they could move, tried to escape to safety. Even where the city was not flooded, the damage was considerable. At top right a car has been crushed by a falling wall. At bottom right survivors are stacked five high in a Hercules aircraft as they are being evacuated for medical treatment.

Dennis
July 10

Emily
July 18

Irene
August 14

Ophelia
September 11

On this spread: There were so many tropical storms and hurricanes in the Atlantic in 2005 that meteorologists had run out of names by mid-October. Tradition dictates that successive storms are given names starting with successive letters of the alphabet. Although difficult letters like X are omitted, there are normally more than enough names available for each storm season. In 2005 the Greek alphabet had to be pressed into use, and when the twenty-third storm of the Atlantic hurricane season formed off the coast of Panama late on October 26, it was dubbed Beta. The storm set a new record for the number of tropical storms to form in the Atlantic during a single year, a record that was broken with the formation of four more in November and December, which also made it the longest storm season on record. Scientists believe there was more energy in the system because the water temperature in the South Atlantic and the Gulf of Mexico was at record highs. It is often heavy rain, causing mud slides, and tidal surges that cause the most destruction. Beta, for example, dumped 16 in. (450 mm) of rain in 24 hours in Nicaragua, even though it was only a category 2 hurricane. These storms, shown above with their names, caused damage over wide areas of the Caribbean and Central America, where for ordinary homeowners it is no longer possible to get storm insurance. Homes now have to be built to be hurricane-proof before they can be insured. Hurricane-force winds from these storms could be felt up to 90 miles (145 km) from their center, and lesser but still storm-force winds reached 230 miles (370 km) away.

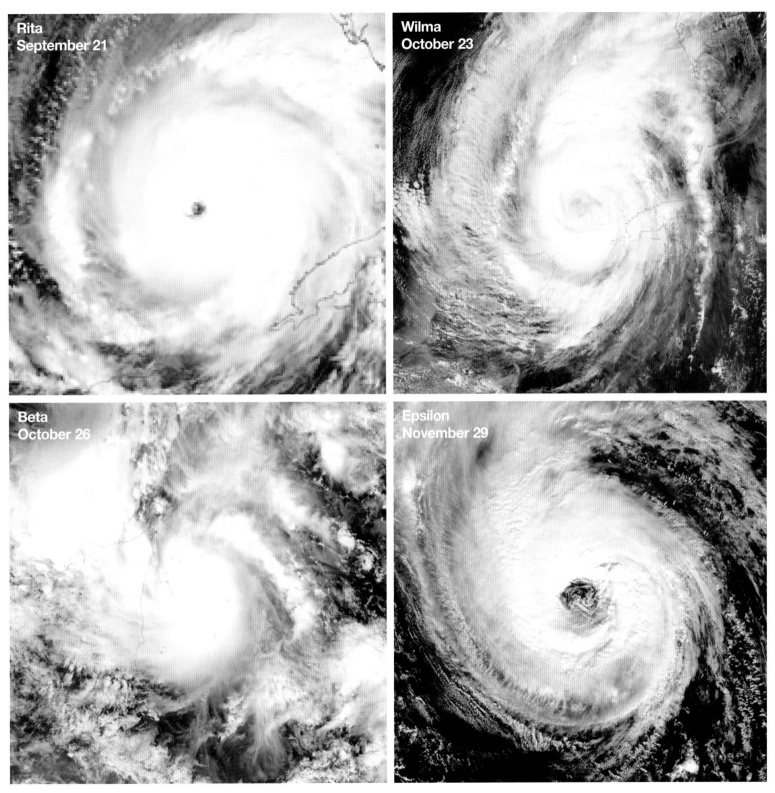

Rita
September 21

Wilma
October 23

Beta
October 26

Epsilon
November 29

Hurricane Dennis *(opposite, top left)* put clouds of rain over most of the southeastern United States well before the storm came ashore. In this image of Dennis taken on July 10, the storm already covers all of Florida, Alabama, Mississippi, and parts of Louisiana. The economic cost of interruptions to normal life and damage by storms is measured in insurance claims, but the true total cost is almost impossible to calculate.

In the days before Hurricane Katrina struck, new data began to appear to remove the uncertainty about the violence of storms. Most predictions showed an increase in high-intensity hurricanes. Professor Kerry Emanuel of the Massachusetts Institute of Technology said that the destructive potential of tropical storms had doubled in the last 30 years. One of the reasons the danger had been underestimated was the failure to take into account the fact that oceans were warming to greater depths. Previously when a storm began, the wind broke up the surface water, bringing up cold water from below. This cooling down drained the energy from the storm. But Professor Emanuel showed that with warmer waters at greater depths, the cooling effect on mixing up the upper layers of water was reduced or lost, allowing the storms to continue to gather strength and last longer.

Almost as an illustration of this new science, Hurricane Katrina was, in the same month, spinning across the Gulf of Mexico toward New Orleans. It had left the Florida coast as a comparatively tame category 1 hurricane. Within a few hours it began gaining energy from the warm sea and reached the fiercest category 5 before weakening to a category 3 and scoring a direct hit on the vulnerable Mississippi coastline. Shortly afterward a paper in the magazine *Science* showed that hurricanes in the most intense 4 and 5 categories had become twice as common over the last 35 years. The overall frequency of tropical storms had remained relatively constant since 1970, the article said, but the number of extreme events had gone up. Ocean surface temperatures have risen by an average of 0.5°C over the same period, almost certainly directly linked to global warming. In 2005 the waters in the Gulf of Mexico were a full 1°C warmer than the long-term average.

By the end of the 2005 hurricane season, when the records were compiled, it was clear that it was not just the intensity of storms that had increased. Significantly, the season had started earlier. There had been more storms in June than ever before, and the season had lasted until December 6, the latest date a hurricane has been recorded in the Atlantic. Hurricane Wilma, in October, also broke the record as the most intense hurricane ever recorded, producing more than 60 in. (1.5 m) of rain. Insurance losses broke all records in the season—about $100 billion, of which $34 billion was as a result of Katrina.

Despite these statistics, which went beyond anything suggested by the predictions of scientists from their own computer models, there remained argument among the experts. However, in 2007 in the new Intergovernmental Panel on Climate Change (IPCC) report, there were new statistics on the frequency of hurricanes. This showed that since the 1970s the number of category 4 and 5 storms, the most violent, had doubled. In the 1970s in each five-year period, there were an average of 45 such storms around the world—in the warm waters of the Atlantic, Pacific, and Indian oceans. In the 1990s to 2005, the five-year average was 90.

It is still possible that these stronger storms are just a natural variation in the climate and not made worse by man-made global warming, but this now seems unlikely. In view of this evidence, it seems odd to outsiders that New Orleans is being rebuilt in the same place. Despite the enormous goodwill toward the victims of the hurricane, particularly the poor, it cannot be sensible to rebuild the suburbs that are below sea level. The desire to keep the center of the city from losing its real character and charm and becoming a theme park is understandable,

On this page: Months after Hurricane Katrina devastated large parts of New Orleans, some suburbs still remained in ruins. The storm struck in August 2005; in March 2006 members of the People Improving Communities through Organizing (PICO) toured the region. They had chastised the U.S. government for delaying billions of dollars in aid for a region known for its historic churches. At the top Gloria Cooper of San Diego was still able to see this damage in the Lower Ninth Ward of the city. At the bottom other members of PICO look at a damaged house in the Gentilly neighborhood.

Opposite: The scale of the damage to a once bustling city was only just becoming apparent two days after Hurricane Katrina struck New Orleans. At this stage thousands of people were unaccounted for, and rescuers had not reached many people who were still stranded in their wrecked homes. Many of the poor were trapped for up to a week by debris and floodwater, having been unable to join the prestorm evacuation because of a lack of transportation.

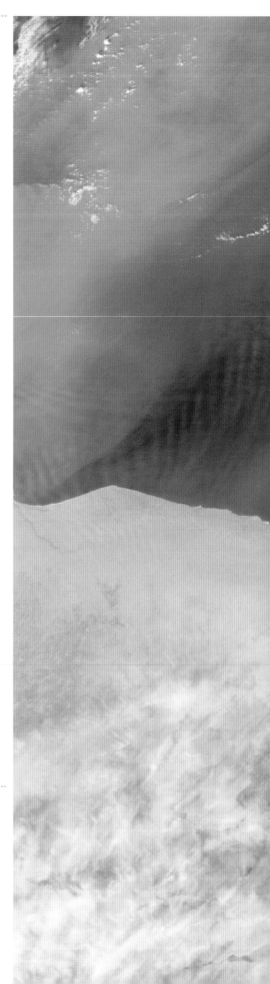

"Canadians are facing risks such as the spread of disease, more drought in the prairies, melting permafrost in the North, longer and more intense heat waves and smog, and rising coastal waters."

—Johanne Gélinas, Commissioner for
Environment and Sustainable Development

but rebuilding the city as it was can only be for the short term. After all, whether the hurricanes are a result of natural variability or not, everyone agrees that another direct hit by a hurricane is a certainty sooner or later. Next time it could be worse. Hurricane Katrina has already destroyed some of the offshore islands that help defend New Orleans from the storm surge, and the sea level is continuously rising.

One other oddity of the season was a hurricane called Vince, which began off the coast of Africa in October, but instead of tracking west toward the Caribbean, it spun off northeast past Madeira. It eventually made landfall in Spain as a tropical depression. This is the first time this kind of storm has been recorded reaching the European mainland. Ironically the large quantities of rain deposited by Vince came as a relief to the Iberian peninsula because it had been suffering the worst drought in southern Europe since records had begun 60 years ago. Even orange trees died in the 42°C heat as irrigation water dried up. This is one of the long-term climate-change predictions that scientists have made—the Sahara desert leaping the Mediterranean. The south of Spain, Portugal, France, all of Malta, Sicily, the toe of Italy, Greece, and parts of countries bordering the Black Sea are all drying out. This is because of the misuse of water resources (for example, for the tourist industry and wasteful irrigation practices, as well as climate change).

Perhaps golf courses are the best, or worst, examples of this unreasonable use of water. Dutch engineers working in Spain on water conservation have worked out that one 18-hole course can use as much water in a year as a town of 10,000 houses. Nine liters of water per square meter per day are needed to keep a golf fairway looking green in Spain. More than

40 new golf courses are planned for the Murcia region alone and as many again for Alicante and Almería. These are three of the driest areas in Spain. Kofi Annan, the former U.N. secretary general, has commented: "In southern Europe lands once green and rich in vegetation are turning barren and brown."

Despite what appears to be a developing battle between the competing interests for depleted water resources in southern Europe, these countries, with their wealth, are better placed than many on other continents. Even so, as a result of their plight, these states drying out on the edge of the Mediterranean are all members of the U.N. Convention on Desertification and Drought. The convention has 179 members, and according to the United Nations, 110 countries are already affected by desertification. The United Nations says the livelihood of one-fifth of the world's population is threatened, and an estimated 135 million people are at risk of being displaced. The worst affected continent is Africa. The United Nations estimates that by 2020, 60 million people will leave the Sahelian region of North Africa if desertification is not halted. Nigeria, with the biggest population in Africa, is losing nearly 1,500 sq. miles (more than 350,000 hectares) of grazing and cropland to deserts each year. Kofi Annan has said that the spread of deserts, environmental degradation, and poverty are all closely linked and must be tackled together.

But Africa is not the only area badly affected. The northeast of Asia has suffered severe dust storms in the last two decades, and some settlements, schools, and even airports have been buried in sand. Iran reported in 2002 that sandstorms had buried 124 villages in the southern province of Sistan-Baluchistan.

Right: A large, swirling mass of dust, visible in the top left portion of the image, is blowing from the Sahara into the Mediterranean Sea. The country on the left is Libya, while the Nile delta of Egypt is on the right. Dust storms are a naturally occurring phenomenon but are made worse by poor agricultural practices that contribute to soil erosion and continued desertification. The storms spread dust over vast areas of ocean, Europe, and even the Amazon, depositing minerals as the dust falls. These can have a fertilizer effect in the sea, stimulating plankton growth and adding nutrients to aid tree growth in the forests. In the air they can also have an effect on rainfall. Scientists are still trying to understand their full impact. Governments and ordinary people are planting trees and creating other barriers with vegetation in an attempt to keep the sand in place.

"Africa is our greatest worry. Many countries are already in difficulties, and we see a pattern emerging. Southern Africa is definitely becoming drier."

—Wulf Killmann, chair of the U.N. Food and Agriculture Organization's climate-change group, commenting on the fact that 34 countries including Ethiopia, Zimbabwe, Eritrea, and Zambia were experiencing drought and food shortages, summer 2005

China has a tree-planting program to try to halt desertification and even push back the frontiers of the sand dunes. Its target is to plant a billion new trees. Parts of North America are also drying out and have suffered dust storms, raising the specter of the dust bowl of the 1930s. So far in the world as a whole, the battle against deserts is being lost. Partly through natural desert spread but also because of overgrazing, poor irrigation practices, and deforestation nearly 25,000 sq. miles (around 6 million hectares) of productive land are lost each year.

Each of the countries involved in the convention has developed a plan to take on the desert conditions in its own territory. So far the most successful schemes have involved local people, usually women, in planting suitable vegetation to hold back the sand. Large-scale plans are often expensive failures because in harsh conditions trees and shrubs need individual attention and watering to survive long enough to take firm root.

As reported above, the United Nations has repeatedly made the link between poverty and land degradation. One of the big aims of the 2002 Earth Summit in Johannesburg was to improve the freshwater supply for the 1.1 billion people without it and provide sanitation for more than 2.4 billion. The lack of care in using water resources and keeping the organic content of soil is linked to the ability of people to grow their own food.

On a small scale, particularly in southern Africa, where these issues have been tackled together, it has been possible to reverse desertification and create thriving self-sufficient communities. These demonstration projects have yet to be adopted on a wide enough scale, but even if they were, it may not be enough. The Hadley Centre for Climate Change, the UK's official center for climate-change research, predicts big increases in extreme and severe droughts across large areas of the world as a result of man-made climate change.

The problem of deserts has also been linked to the loss of forests. It has certainly been a major factor in land degradation. In the Amazon the destruction of once virgin forests is expected to lead directly to the creation of deserts. Over the planet as a whole, it is an amazing fact that already more than half the original forest cover has been destroyed, although much of this was in Europe and North America. Vast areas have disappeared in the tropics in the last three decades. It is possible to play with statistics to show how quickly tree cover is disappearing. In the 1990s the rate of forest destruction was 62,200 sq. miles (16.1 million hectares) a year. Put another way, that is 4.2% of the world's virgin forests in a single decade, or in tabloid terms 33 football fields a minute, or each year an area the size of Portugal. Whichever way it is interpreted, this level of loss is clearly unsustainable.

As mentioned earlier, the fight to save the world's forests has been a long and unsuccessful one, although politically it appears that prospects are improving. Forests are vital, from the view of biodiversity as well as climate. More species inhabit the tropical forests than any other habitat on Earth. Many of their secrets are still undiscovered, along with the potential benefits in food sources and drugs. They are all being lost before they are even discovered. Forests are worth saving for that reason alone, but trees are vital for many reasons to do with storing carbon, managing water resources, and climate change. Trees must capture carbon dioxide in order to grow, so the cutting down of the forests is releasing huge quantities back into the

(continued on page 199)

195

Opposite, top: Lionesses cross the dry Ewaso Ngiro riverbed in Kenya's Samburu game reserve in January 2006. The lion, known as the "king of the beasts," is losing its habitat so fast that it may disappear from the wild. Increasing conflict between humans and the big game of Africa means that the territory where these animals can thrive is being squeezed.

Opposite, bottom: Wildebeest, ever alert for marauding lions, are framed against an African sunset in what is a reminder of how majestic Africa's wildlife can be. Droughts, believed to be symptoms of global warming, and human pressure for land are cutting into the traditional migration area for the once huge herds.

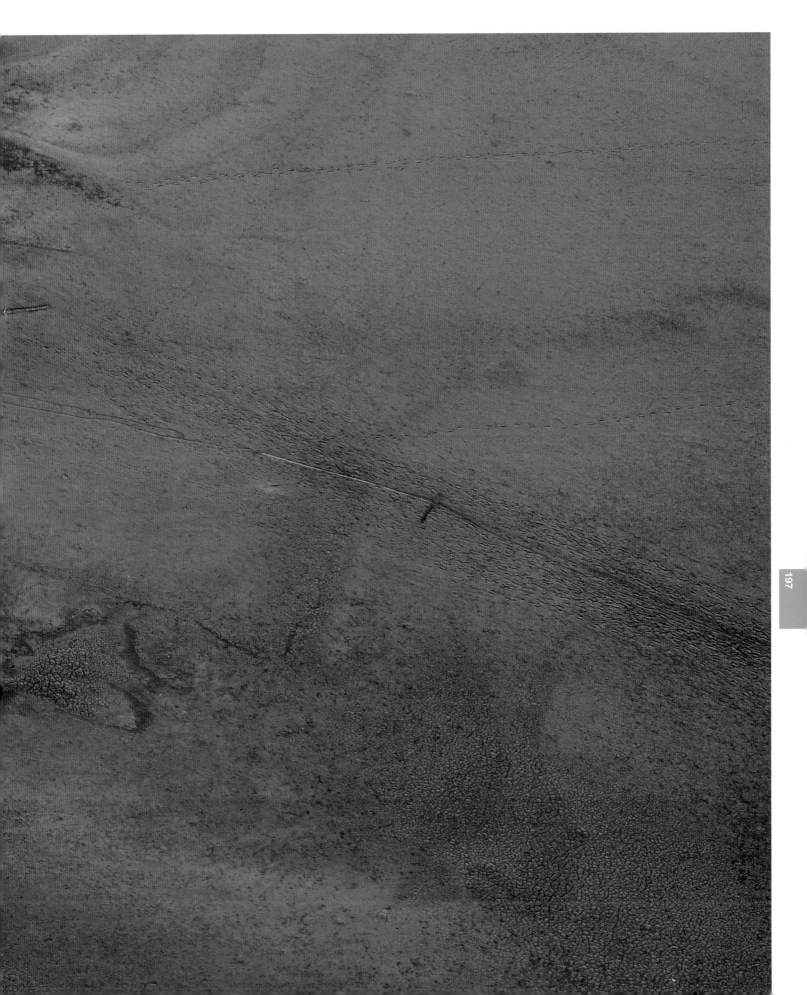

Previous spread: The drought that hit the Amazon basin in 2005 stopped the flow of some of its tributaries, stranding river steamers, which are the main means of transportation in a region without roads. Scientists fear that this drought may be the start of the drying out of the region that will lead to the destruction of large parts of the forest, which have been described as the lungs of the world. The extra warmth of the sea in the Gulf of Mexico is believed to be part of the reason, along with the reduction in the size of the forest through fires and land clearance. Many of the bigger trees were damaged so fiercely by the drought that they are dying.

atmosphere. Some estimates are that around 25% of the extra carbon dioxide being released into the air by man comes from the destruction of forests. In addition, all over the world where forests have been cut down, particularly in mountainous areas, devastating floods have resulted. The Chinese government has ordered the reforestation of its mountainous regions to regulate the flow of its rivers, so trees are both holding back floods and halting spreading deserts. They are also used as an effective break against avalanches, and on the coast as protection against erosion, sea-level rise, and storm surges.

Trees effectively act as a giant sponge to soak up the rain and gradually release the water into streams. Even in Europe, where winter and spring floods are becoming an annual feature, it has been found that trees planted on the edge of tributary streams can hold back and lessen the impact of pulses of floodwater that are causing so much damage.

Trees also evaporate water through their leaves, creating more water vapor, and as a result, forests release enough moisture to form new clouds and cause repeat rainfall farther downwind. Forests are creators of their own weather in another sense. For one thing, they have an uneven surface compared with grassland, and they also absorb more heat from the sun. This combination makes turbulent damp air rise, stimulating further rain. In well-wooded tropical regions this often leads to sharp showers in the afternoon after a day of sunshine. If grassland replaces forests, there is a considerable reduction in rainfall.

Computer experiments show that if the Amazonian rain forest were cut down and replaced by grassland, then rainfall in the region would fall by as much as 70%. There was speculation after the Amazon drought of 2005 that the loss of vast tracts of rain forest in the last 50 years was already having a serious effect on the region. By midsummer the greatest river on Earth had already dropped 6 ft. (1.8 m) below normal level to less than 49 ft. (15 m) deep. Rainfall was less than 65% of the July average. But as with all averages, these figures disguise the extremes. Some tributaries, normally hundreds of meters across, dried out completely. Whole villages that relied on fishing for trade and survival were abandoned, and river boats, the only form of transport in some places, were grounded and left unmanned in midstream.

It was the worst drought anyone could remember. However, records in this region go back only 40 years, which is not long enough to tell whether it happened before. But the drying of the rivers came after five years of low rainfall, a potentially disturbing trend. Environmental groups immediately blamed at least part of the problem on deforestation altering local rainfall patterns. Scientists also believe there may be a link between the drought and the extra warm waters in the southwest Atlantic. The warm sea drew the moist air from the Amazon and provided energy that contributed to the record hurricane season in the Caribbean. Either way, for the river communities of the Amazon, one of the devastating problems was that even where there was water, millions of fish died from lack of oxygen. Warm water has less oxygen. In some places rotting fish floated across the whole surface of stagnant lakes and pools. Again time will tell whether this was a one-of-a-kind extreme event or the beginning of a pattern. True total cost is impossible to calculate.

Opposite and above: The lakes of the Amazon basin and the rivers are interconnected and teeming with wildlife in one of the hottest and steamiest regions of the Earth. Drought is not unknown, but very rare and on the scale of 2005 has catastrophic consequences for the wildlife and human population. Here a horse trots across the dry bed of Brazil's Curuai Lake. The fish congregated in ever smaller patches of water and streams until lack of water flow led to many drying out altogether. Even where some water remained, a lack of oxygen meant that fish died in the millions, leaving this fisherman with no harvest even when the rains returned.

Fiddling While the Globe Burns

Previous spread: Demonstrators depicting President George W. Bush as fiddling while the Earth burns are comparing him to Emperor Nero, who is said to have done the same when Rome was burning. The environmental group Energy Action staged their protest outside the U.N. Climate Change Conference in Montreal on December 5, 2005, calling on the United States to rejoin the Kyoto Protocol. They argued that the administration was out of touch with its citizens.

Above: Trees are particularly vulnerable to climate change because they rely on conditions remaining the same for many years so that they can grow, fruit, and flourish. They live a long time and generally drop seeds close by, so they also migrate more slowly than any other plant. Here in Portugal a tree dies after succumbing to drought.

Above: Drought is an obvious danger to forests because dry trees make it much easier for fires to break out. Here a villager tries the impossible task of extinguishing a forest fire in Abrantes, central Portugal, on August 22, 2005. These fires raged across the country for several weeks after a two-year drought left the whole region tinder dry.

UN HQ / New York

YURIKO WANGARI MAATHAI GINÉS GONZÁLEZ GARCÍA MASAO NAKAYAMA JOKE WALL

Left: Kofi Annan, then U.N. secretary general, speaks via a videotaped message during the ceremony in Kyoto city to mark the entry into force of the Kyoto Protocol. Mr. Annan, who was at the U.N. headquarters in New York, had been a vocal supporter of the protocol during the tortuous seven years of heated negotiations needed to reach this moment. It was celebrated by its backers as a lifeline for the planet but rejected as an economic straitjacket by the United States and Australia. On the podium are Wangari Maathai of Kenya, winner of the 2004 Nobel Peace Prize; Gínes González García, Argentine environment and health minister; Masao Nakayama, Micronesia's representative at the U.N.; and Joke Waller-Hunter, then executive secretary of the United Nations Framework Convention on Climate Change.

> "Climate change is a global challenge that demands a global response. Yet there are nations that resist, voices that attempt to diminish the urgency or dismiss the science, or declare, either in word or indifference, that this is not our problem to solve. Well, let me tell you, it is our problem to solve.... To the reticent nations, including the United States, I say this: There is such a thing as a global conscience."

—Former Canadian prime minister Paul Martin, quoted in the *L.A. Times*

There were tears and congratulations in the corridors and in the conference hall at the end of the Montreal climate talks in December 2005. It was, as these moments often are, described as historic. But will historians record this mammoth conference as a success or a failure, or something in between?

If mankind is to survive the potential catastrophe he has created by changing the world's climate, the foundations of a workable plan to adapt and cut emissions should surely have been laid in Montreal. An attempt was made, but will it be enough? The elated and sometimes tearful reaction to the agreement at the end of two weeks of bargaining was partly because of the relief of exhausted delegates after the conference had teetered on the edge of disaster several times. From the beginning the toughest task of the delegates of the 189 countries represented was to re-engage the American delegation in the process of combating climate change. For many the fact that the administration of George W. Bush was even prepared to talk in the future about how to reduce emissions was a victory. For others it was fiddling while the Earth burned.

It is important to examine what happened at the conference for two reasons. The first is that the 1992 Climate Change Convention and its 1997 addition, the Kyoto Protocol, are the only credible international mechanisms for reducing greenhouse gases. As has been discussed earlier in this book (in the section on the history of climate change), the 1992 convention's "ultimate objective" was the "stabilization of greenhouse gas concentrations in the atmosphere that would prevent dangerous anthropogenic [man-made] interference with the climate system." In the 13 years between then and Montreal, the science community had overwhelmingly reached the view that the human race was already responsible for dangerous interference with the climate and was every day making the situation worse. It would appear then that the time for further discussions of what to do should have been over. The politicians at the conference, all of whom helped run countries that had signed

up to the convention, were therefore obliged to take action on policies and actions to solve the problem.

The Kyoto Protocol, the add-on to the original convention, provided the major mechanism for taking these steps by giving all industrial countries legally binding targets that must be met by 2012. Although all 36 industrial countries originally signed up to this deal, George W. Bush repudiated America's target and his country's involvement as soon as he was elected. Only Australia followed his lead.

Three years before, in 1997 when the protocol was negotiated in Kyoto, President Bill Clinton had agreed to reduce America's greenhouse gas emissions by 7% by 2012. President Clinton was unable to get the U.S. Senate to ratify the 1997 deal because many Americans believed it would make their industry uncompetitive with countries that had no restrictions on their greenhouse-gas emissions. For its part the Australian government was trying to protect both its own energy-intensive industries and its large exports of coal.

To understand what happened at Montreal, it is important to realize that these previous political maneuverings had created two conferences in one. The first involved all those countries that had signed the original convention: 189 at the start of Montreal, virtually all the world's nations. The second involved only the more select number who had also ratified the Kyoto Protocol. These came to a total of 156 of the 189. This second group, of course, included the 34 developed countries, which had accepted legally binding greenhouse-gas reduction targets, and excluded the two countries that had repudiated the agreement, the United States and Australia. The other members of the Kyoto club

(continued on page 209)

Opposite, right: Still fighting political progress on climate change every step of the way, Paula Dobriansky, U.S. undersecretary, democracy and global affairs, puts forth the White House point of view at a news conference during the U.N. Climate Change Conference in Montreal on December 7, 2005. Next to her is Harlan Watson, senior climate negotiator. The United States did eventually agree to talk about taking action to curb climate change when the Kyoto agreement expires in 2012.

Above: Floods are nothing new in the United States, but just as everywhere else, the number of extreme weather events is growing. This is Conroe, Texas, George W. Bush's home state and one of the country's driest, where the headquarters of America's oil industry is also located. Here in 1999 a grandfather carries his 11-month-old granddaughter to safety through rising floodwaters while his wife holds their dog's leash. They were evacuating their home.

"It's obvious we can't ignore this problem any longer. Locally and nationally, we cannot wait to see how bad it gets. We need to act now."

—Greg Nickels, mayor of Seattle

Above: Rescuers save a one-year-old child from floods near Guayama, Puerto Rico, on September 10, 1996, when Hurricane Hortense struck Central America. Hortense brought torrential rain through the Lesser Antilles, Puerto Rico, and the Dominican Republic. It caused 39 deaths and $158 million worth of damage.

"The governments of the world have tarried long enough, and the United States is scarcely without doubt the greatest culprit among them."

—Walter Cronkite

are developing countries who are in support, including, of course, all of the small island states that face extinction because of sea-level rise.

In legal terms the Kyoto Protocol had only come into force on February 16, 2005, after Russia had finally ratified. The Montreal meeting was therefore the first opportunity for all of the participants to meet as a kind of club, and so the conference was officially the first Meeting of the Parties to the Kyoto Protocol, or in conference jargon MOP 1.

At the same time as the Kyoto partners were meeting to finalize the details of how the 1997 protocol would operate, a second parallel conference of all 189 countries that had ratified the original convention was taking place. This included the United States and Australia and was known as the 11th meeting of the signature nations of the 1992 Climate Change Convention. As described in an earlier chapter, the first of these conferences had been held in Berlin in 1995. Under the terms of the treaty, there has to be at least one conference of the parties, or COP, each year ever after. So the meeting was both the first Meeting of the Parties for the Kyoto Protocol, or MOP 1, and COP 11. Both had important business to transact.

This distinction between the two conferences may sound pedantic but is in fact politically vital because for the first time since the original convention came into force in 1992, the Americans were excluded from part of the discussions. The implementation of the Kyoto Protocol was a matter only for those nations that had ratified it. In Montreal the most powerful nation in the world was in effect an outsider, not a member of the club, and unable to influence its proceedings.

This change must be seen in the context of the previous seven years, when Americans had been free to take part in all of the discussions, because until 2005 the Kyoto Protocol had not been in legal force. For at least the last five of those years, since George W. Bush had become president elect, the American delegations had done their best to undermine and destroy the process at each annual round of talks.

After first confidently expecting the Kyoto Protocol to collapse entirely in 2000 when the United States pulled out, and having been disappointed, United States delegations had subsequently thrown every spanner they could into the works of the Kyoto Protocol. The fact that it still did not collapse and the rest of the world decided to proceed to tackle climate change without the United States was undoubtedly a constant irritant to the White House. It would be wrong to say the United States was alone in its endeavor and did not have allies in this attempt at wrecking the protocol. Oil-rich Saudi Arabia and coal-rich Australia were constant in their support of the United States, creating delays and obstructions at every opportunity. Other countries, in difficulties with their own targets or seeking deals on technology or carbon trading, also covertly supported the United States from time to time. They were happy to see progress slowed down but equally cheerful about the United States getting the blame, taking care not to allow their domestic audience to see them playing an anti-climate hand.

There can be no doubt, however, that the United States was the main player and ringleader and its fossil-fuel industry the principal funder of anti-climate lobbyists and the greatest single influence on the White House. At Montreal, for the first time, the United States could not play its traditional obstructive role. It was in effect out

Opposite: U.S. president George W. Bush walks hand in hand with Saudi Arabia's Crown Prince Abdullah among Texas bluebonnet wildflowers on his ranch in Crawford, Texas, April 25, 2005. During his meeting with Crown Prince Abdullah, President Bush praised the kingdom's efforts to fight terrorism and sought the prince's help in bringing down the price of oil. The two men have done their best to slow down international efforts to fight climate change. Saudi Arabia has the world's largest oil reserves and claims that combating climate change would damage its economy. At all climate talks since the Kyoto agreement was signed in 1997, the Saudi Arabian delegates have demanded compensation from developed countries for the loss of oil sales.

Above: Everywhere the U.S. president goes he attracts demonstrators, but an increasing number of them, as shown here during a visit to Brussels in 2001, are concerned about climate change. This was shortly after he had repudiated the Kyoto Protocol and said the United States would never ratify it.

in the cold, outside the new club of nations. This had a significant psychological effect on proceedings, which moved much faster as a result. In 1997 when the Kyoto Protocol with its set of targets for industrial countries had been agreed to in principle, there had been a whole series of complex problems to sort out in the small print. These were vital to the successful operation of the agreement. Among them were two methods of transferring technology to poorer countries and saving large quantities of emissions in the process.

The argument for these two schemes was simple. The climate did not care where the reductions in emissions were made as long as they were made. The protocol invented two ways technology transfer could take place between nations to the benefit of both countries involved; one was called Joint Implementation, the other a clean development mechanism.

The first relied on large sophisticated countries like Germany and Japan, which had already done a great deal to make their power plants efficient, investing in other industrialized countries to bring them up to similar standards. Older plants in former Soviet-bloc countries like the Ukraine, Romania, and Bulgaria were incredibly polluting, but the governments concerned either did not have the money, or the technology, or both to update the plants themselves. Under the treaty it was made possible for rich countries with good technology to pay for upgrades of equipment in less efficient ones and so claim "carbon credits" for cutting down greenhouse gases. Remember the argument was simply that the climate did not care where the savings were made as long as they were made.

The second process, the clean development mechanism, had the same motive, saving the

climate, and it was also a way of transferring technology to the developing world. The idea was that developing countries that wished to undergo an industrial revolution themselves, to create wealth and employment, should not go down the same dirty route as Europe and North America had gone. Instead, new technologies, like solar, wind power, and small-scale hydropower, would be installed in developing countries to provide electricity, leapfrogging over a century of dirty fossil-fuel development straight into the 21st century. Those countries providing the technologies would be able to claim carbon credits that would count toward their own domestic targets of greenhouse-gas reductions. This was a way of aiding the developing world without having to impose potentially unpopular draconian restrictions on voters at home. Some of the schemes were controversial and some technologies excluded. For example, nuclear power, while a low-carbon emitter, was not regarded as a clean technology. Large dams were also excluded, and there was a great deal of debate about reforestation projects. How much credit should be given for planting trees to capture carbon and for making changes in farming practices to prevent carbon from being released from the soil in the first place are a matter of scientific dispute and, as a result, politically controversial.

One other difficult area that had to be settled was the penalties for failing to meet targets set down in the protocol. This is tricky because without sanctions on nation states, which would have to be noticeable, politically embarrassing, and inconvenient, how can any legally binding international agreement be made to work? How could they be organized so that they did not also appear to infringe on national sovereignty, a potentially touchy subject? Successful negotiations on sanctions had taken place at

Above and opposite: The old technologies and the new. Ratcliffe-on-Soar, a coal-fired power station near Nottingham, England, was built next to the coal fields that once powered Britain's industrial revolution. The power station is one of Britain's biggest, but coal produces more carbon dioxide for each unit of electricity generated than any other fuel. This global warming gas needs to be captured and stored underground if power stations like this one are to continue in operation without further damaging the planet. The lower picture above shows one of the potential solutions: solar panels. These are in Pellworm, Germany, and produce power directly for use in the building on which they stand without any carbon emissions. *Opposite top:* The 100-year-old Iljitsh oil field in Azerbaijan on the Caspian Sea is one of the oldest oilfields in the world, and one of the most polluted. It is a forest of rigs, some disused and some still operating but leaking, standing in lakes of oil more than a meter deep. This is in sharp contrast to the clean lines of the offshore wind farm at Middelgrunden, Denmark, (opposite, bottom) which, once built, produces carbon-free electricity.

"Those least able to cope and least responsible
for greenhouse gases that cause global warming
are most affected. Herein lies an enormous
global ethical challenge."

—Professor Jonathan Patz on behalf of the World Health Organization,
which estimated that 150,000 people a year already die and 5 million
are made severely ill by climate change, November 17, 2005

"We should be treating, I think, the whole issue of climate change and global warming with a far greater degree of priority than I think is happening now."

—Prince Charles, BBC interview October 2005

earlier meetings of protocol parties, but they remained to be brought into force in Montreal. The failure to reach the target will not become politically embarrassing until 2012, but by then the world might be far more concerned with climate change than it is now.

The working of the sanctions agreement will be assessed and applied by an expert panel. This is already set up and is monitoring progress toward targets. In 2012 it will ultimately judge whether each country has reached its commitment with a review of its final annual emissions inventory. If the panel concludes that a country has failed to meet the requirements, the government concerned will be given 100 days to buy carbon credits or use any of the other bargaining or compliance mechanisms allowed for in the protocol. "If, at the end of this period, a party's emissions are still greater than its assigned amount, it must make up the difference in the second commitment period, plus a penalty of 30%," says the convention rule book.

It must also develop within the following three months a "compliance action plan" to show how it is going to meet its target in the next commit-ment period. Although these can hardly be described as draconian measures, they had taken years to negotiate, and there were fears before Montreal that there would be further holdups, objections, and obstructions, as there had been at every previous meeting. Remarkably nothing substantial was raised. Without the Americans present observers remarked that there was "an optimistic atmosphere—a desire to get on with it," as one senior delegate put it. The MOP 1 meeting was concluded successfully, and the Kyoto Protocol, at last fully in force after Montreal, became the first real test of man's ability to solve the climate problem. There remained at the end of the

meeting seven more years for participants to put policies in place to reach the targets they had agreed to in 1997 in Kyoto. Some of these had been agreed to by previous leaders from different parties, but unlike the United States and Australia no other leader had gone back on his country's pledges. A few were already on target to meet the promised reductions, but most were not, and they accepted the need for a big new push to reduce emissions.

In most countries, when the crunch year of 2012 comes, the governments of 1997 and 2005 will be part of history, and a new leader will have to face the blame and the penalty. Delegates real-ized this danger, and the treaty has sensibly built in an annual reporting mechanism so that each government reports progress in reaching its targets. As each report comes in, it is painfully obvious which countries are failing. Under the treaty, when they are wide of their targets, they have to offer a program of how to get back on track. In 2005 many countries whose economies had been growing since 1990 were already struggling to meet targets.

But back to Montreal. The details of the Kyoto Protocol having been settled, the next big step was to conclude the second parallel part of the conference, COP 11. This, of course, included the United States. The principal task of this part of the meeting was to decide what was the next step—beyond the end of the Kyoto agreement in 2012—to reduce emissions further. The treaty had built into it the requirement that by 2006 parties to the Climate Change Convention would be discussing what to do beyond 2012. Remembering that the Kyoto agreement overall only reduced greenhouse-gas emissions from the industrialized world by 5.2% and scientists advised that cuts from the entire world needed to be between 60% and 80%, this was quite a tall order. If cuts of this magnitude are ever to

Left: The best intentions of governments can be laid waste by extreme weather events. Iran is suffering from desertification, and the oil- and gas-rich country is attempting preventive measures. Here pine trees planted in a government reforestation plan lie flattened by floodwaters in the Minoodasht area of northeastern Iran in August 2001. Devastating storms in northeastern Iran killed some 300 people, following what is thought to have been the region's worst flooding in 200 years.

"We're seeing that climate change is about more than a few unseasonably mild winters or hot summers. It's about the chain of natural catastrophes and devastating weather patterns that global warming is beginning to set off around the world—the frequency and intensity of which are breaking records thousands of years old."

—Senator Barack Obama

Above: Some American politicians refuse to accept the term global warming and insist that it must be called climate change. In the sense that the extreme weather events can mean sudden cold as well as extra heat and bigger storms, they are right. Here pedestrians make their way through snow and slush in New York in February 2006. The thigh-high snow was said to be a record in the city and caused the cancellation of flights and delays to trains across the northeastern United States.

be achieved, then the whole world needs to be involved.

Although it would be possible for the Kyoto parties, already adopting new technologies and measures to reduce emissions, to continue with the task, they could not alone reach a 60% reduction target for the planet. Without the United States, which produces 25% of the world's carbon dioxide on its own, no agreement could save the Earth from uncontrollable climate change. In addition, the developing nations, including the new industrial giant China, the ever growing India, and rapidly advancing countries like South Africa, Brazil, and Mexico, also needed to be brought into any agreement. In fact, it became clear in 2007 that China would soon become the world's biggest polluter, overtaking the United States well before 2010.

But all developing countries are either planning or hoping for rapid industrial advances to fulfill the aspirations of their people, which implies increasing emissions of carbon dioxide and other potent industrial greenhouse gases, including nitrous oxide and methane. It was the "unfair" competition from these emerging economies that the United States used as an excuse for pulling out of the Kyoto Protocol in the first place. It had therefore become clear well before Montreal that all nations of the world must be brought into the process of reducing emissions and must accept their responsibilities to help save the climate. It does not take much imagination to see how difficult this is going to be, but this is what the second part of the Montreal meeting was about: how to develop international agreements to reduce greenhouse gases after 2012. It was just talking about this idea that led the Montreal talks to the brink of disaster. There are some great issues of principle here that divide nations, and blocs of

nations. First of all, as has been previously mentioned, it is the developing world's view that the old industrial nations, particularly Europe, Japan, and the United States, are responsible for most historic greenhouse-gas emissions. They therefore have the main responsibility for solving the problem. Through years of talks, the constant refrain of the developing nations has been that unless the developed countries, which have already grown prosperous by using the atmosphere as a dumping ground for their pollution, make a serious effort to reduce their emissions, then they, the developing nations, cannot be expected to take part. At a minimum they wanted to see these developed countries (including the United States) cut their emissions as part of the Kyoto Protocol agreement before discussing action in the next decade to 2020.

The second major stumbling block, also apparent before Montreal, is the refusal of the United States to consider in any future period the Kyoto formula of giving each country a target to reach and a timetable by which to achieve it. The majority of countries believe that only by legally binding agreements of this kind can they head off the danger of runaway climate change. The U.S. rejects this notion. The position of the world's biggest single polluter is that new technology, either a single magic bullet or a series of them, will be found to solve the problem. Against this background of fundamental disagreements, it is hardly surprising that many people believed that the Montreal talks would collapse in disarray.

But there is a curious momentum to international meetings of this kind. In a world where everyone believes that there is a looming crisis with the climate, no politician, however powerful, wants to be remembered as the man who destroyed the talks aimed at saving the

(continued on page 218)

Above: A rare sight anywhere, ice on an orange tree bearing fruit. Although, unlike other fruit, oranges need cool weather to ripen, they do not need freezing temperatures. This scene, classed as an extreme weather event, was in Hillsborough County, Florida, in January 2000.

Above: Tornadoes have been recorded for centuries, and their destructive power is well documented. The question remains whether with more energy in the atmosphere, in the form of more heat producing more water vapor, the storms are getting worse. Modern technology is certainly making them easier to monitor. This rather grainy picture shows a tornado in Oklahoma in 1973 in its early stages of formation. This image, captured on Doppler radar, is in the National Severe Storms Laboratory (NSSL) collection. Scientists are still assessing whether tornadoes are getting larger, more frequent, and causing more damage.

Above: This is one of the best images yet recorded of a waterspout, a type of tornado that occurs over water. Waterspouts are spinning columns of rising moist air that typically form over warm water. Waterspouts can be as dangerous as tornadoes and can feature wind speeds over 125 m.p.h. (200 kph), a frightening prospect for a small boat that may be in their path. Many waterspouts form away from thunderstorms and even during relatively fair weather. This one was seen off the Florida Keys, arguably the hottest spot for waterspouts in the world. Hundreds form each year over the increasingly warm sea and in the humid atmosphere. Some people speculate that these waterspouts are responsible for the many unexplained losses of ships and aircraft recorded in the Bermuda Triangle region of the Atlantic Ocean.

planet. Perhaps just as important, in the run-up to Montreal, it had become increasingly obvious that the threat was no longer some distant theory and that time was running out.

Another important change had gradually emerged over the previous year. While China is still continuing massive industrial growth with the opening of a new coal-fired power station every week, the government is at the same time seriously concerned about the effects of climate change. Its own environmental crisis—power and water shortages, pollution, and the spread of deserts—are threatening its prosperity, its ability to grow food, and the welfare of its people.

China was not the only developing country where attitudes had changed. Some of the poorest countries in the world have realized that for them the climate-change policies of the richest countries provide a potential opportunity. There is already provision in the Kyoto Protocol under the clean development mechanism for rich countries to claim credit by planting trees in developing countries. One of the little reported but significant developments at Montreal was the proposal by two poor countries, Papua, New Guinea, and Costa Rica, that they should receive financial compensation for keeping their forests standing. It provided a new slant on one of the great unresolved issues of the last 20 years: how to keep the world's remaining tropical forests from being cut down. The nations that contain the forests are among the poorest, and their single greatest asset is the timber in their forests.

As far back as the Earth Summit of 1992, it was forcefully pointed out to the developed world that they could not dictate to others the fate of their forests. It was up to each country to decide what to do with its own natural resources. After

Top: Everywhere the world's forests are under pressure. Usually it is the poorest people who move into the forest to hunt for food and to find a plentiful clean water supply. On the edge of Bach Ma national park, Vietnam, this family uses the forest as a resource for fuel; it needs about 20 bundles of wood a month for a family of six.

Bottom: Elsewhere virgin forest is being cut down for palm oil plantations and other cash crops, threatening the last stretches of forest occupied by rare animals like the orangutan. Here in Malaysia, as a result of deforestation, a three-year-old orangutan has had to be rescued and survives at Sepilok sanctuary.

"If we don't do anything more than what we are doing now, greenhouse gas emissions will rise by 50% by 2030, whereas science tells us that we need to reduce them by at least 50%... our Montreal marathon is over, but we still have a long road ahead of us."

—Stéphane Dion, then Canada's environment minister, in his closing address as president of the U.N. climate-change conference, Montreal, December 2005

all, Europe was once a giant forest, but sadly Europeans had long ago chopped it down. What right did Europe have to tell the rest of the planet how to manage its trees?

But in Montreal this argument took a new tack. Clearly, keeping a forest standing rather than cashing in just once on the timber it contains is a sensible long-term strategy. There is a lot of potential for forest products like nuts and the medicinal use of rare plants, as well as selective use of valuable timber. And a new development these days is in green tourism. Rich Westerners will pay money simply to see such tropical forests and for a chance to observe some of the rare animals they contain.

So 10 countries, headed by Costa Rica and Papua, New Guinea, and styling themselves the Coalition for Rainforest Nations, came up with a surprise package in Montreal. They proposed that poor nations like themselves should be brought into the climate talks as partners. Their contribution to keeping the climate safe would be to maintain their forest cover. It would provide rich donor countries with an opportunity to gain or buy carbon credits by providing aid to keep the forests standing. The donors would also be able to use discoveries of new plants in the forests to develop drugs and for other uses. The credits would be granted on the basis of carbon dioxide saved from going into the atmosphere, and they would be used toward the donors' own domestic targets. As Carlos Manuel Rodriguez, Costa Rica's environment minister, put it: "If we do not recognize the value of these forests, they will be cut."

Apart from being a rather good strategy to save threatened forests, this idea also had an interesting political dimension. It implied that some of the poorest developing countries, with no blame for causing climate change, were

prepared to be brought into the process and at least accept one important target—not to cut their forests down. It was the first time that developing countries had accepted such an idea.

Toward the end of the conference, the new mood of developing countries became apparent when a maverick Russian suggestion failed to get any support. The Russians proposed, for reasons that never became clear, that developing countries should not be asked to accept targets for cutting greenhouse-gas emissions or timetables to do so. Observers expected that at least some developing countries would be in favor of such a free ride. None were, and the Russian proposal was withdrawn.

The new mood among developing countries is partly spurred on by the fact that the dangers of climate change are already obvious in many of them, and not just in the low-lying island states. It is the poor that are always the worst affected, an idea that has been reinforced several times since the Montreal meeting, including the Stern Review and the three most recent Intergovernmental Panel on Climate Change reports in 2007.

None of that solves the problem of how to engage the United States in the process. The rejection of the idea of targets and timetables to cut greenhouse-gas emissions, which the rest of the world sees as the way forward to deal with the crisis, leaves such a gulf between the two sides that it is hard to see how it can be bridged.

To prove the point, toward the end of the Montreal conference, in a moment of high drama, the American delegation walked out. The talks were on the point of collapse, and across the world the news made headlines. One British

Above: The 160 ft. (50 m) long mosaic of 3,000 tiles bearing climate change messages painted by people from all over the world, which was built directly opposite the Palais de Congrès in Montreal, the venue for the 2005 U.N. climate talks in the city. It was organized and assembled by the international environment group Friends of the Earth.

"There is no longer any serious doubt that climate change is real, accelerating and caused by human activities."

—Bill Clinton, former US president, speaking to the climate conference in Montreal, Canada, December 9, 2005

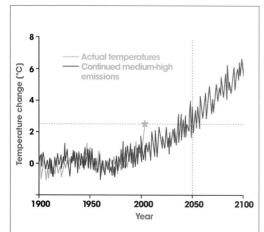

The heat wave in Europe that killed thousands of people in 2003 was unprecedented in recorded history. Scientists called it a once-in-1,000-year event. The asterisk marks how much above average the temperature was for more than two weeks in 2003. This graph shows the continuation of expected global warming after some attempt to reduce greenhouse-gas emissions but still leaving them at medium-high levels. By 2050 it is clear that the temperatures of 2003 will be normal in southern Europe. By the 2060s the exceptional heat of 2003 would be regarded as a cool summer. If the effect of pollution from traffic and industry, which blocks some of the sun's rays from reaching the Earth, were removed, this kind of summer would even now be occurring more than once a decade on average.

SOURCE: Hadley Centre for Climate Prediction and Research.

newspaper summed up: "America alone against the rest of the world." But as I said, no one wants the blame for wrecking an important international conference. The White House appears to have been sensitive to the hostile reception that both domestic and world media gave the walkout. In the wake of Hurricane Katrina and the worst storm season on record, the American press, for the first time in recent history, was covering the climate talks in some detail. Within hours of the United States pulling out of the talks, former president Bill Clinton spoke to the conference. It was a brilliant speech, and he didn't pull any punches. His dismay at the U.S. administration's action was widely reported across the United States.

So just as suddenly as they had walked out, the U.S. delegation was back in. Negotiations resumed and eventually, to the cheers and tears that began this chapter, a deal was done. But beyond the obvious relief of those who had struggled through long days and nights to reach a deal, what had been achieved? Essentially, every nation present—including the United States—had agreed to further talks about what to do to reduce greenhouse-gas emissions after the Kyoto Protocol expires in 2012. As one of the detractors of the agreement put it, everyone had agreed to have talks about talks. That's not much of an advance when the planet is said to be in such danger.

But let us go back to the thought that was uppermost in everyone's mind at the beginning of the talks. The major problem since President Bush was elected had been how to get America back into the international process for combating climate change. At the end of the Montreal conference, however small a victory it might seem, the U.S. had agreed at least to talk about what to do beyond 2012 when the Kyoto Protocol expires, even if not an inch of ground

had been conceded on what that might entail. In fact, in the final conference statement, the Bush administration insisted on a get-out clause, which specifically excluded the United States from being tied to any targets to reduce emissions or timetables in which they must be achieved.

Within a few days of the talks ending, the gloss put on the agreement by politicians and some relieved environment groups was already being questioned. Britain's *New Scientist* magazine, read worldwide by scientists who keep up with the issue of climate change, summed up: "In the cold light of day we have to ask what exactly was achieved. The answer looks like little more than an agreement to carry on talking—and even that is hedged in places by promises to talk about very little that is meaningful." The magazine pointed out that while politicians talked, "every square meter of the planet's surface is absorbing about 1 watt more heat than it can release into space. That may be only slightly more than the power of a Christmas tree light bulb. But it matters."

The magazine concludes that the Montreal agreement had done too little to quiet the fears repeatedly expressed by senior scientists, including George Bush's top climate modeler Jim Hansen. The director of NASA's Goddard Institute for Space Studies had predicted, even as the politicians in Montreal were meeting to fudge the final communiqué, that at most the human race had 10 years to make drastic cuts in emissions or face runaway climate change.

So despite the Montreal talks being hailed as a success, it is clear that the gap between the scientists' recommendations for urgent action and the politicians' lazy progress continued to widen. The mood in the United States following

Opposite: Just after the U.S. delegates walked out of the Montreal climate-change talks in December 2005, former president Bill Clinton rose to speak to the delegates. Because the talks were held just across the border, many more U.S. journalists than would normally have covered the talks were there, and the powerful Clinton speech criticizing his successor in the White House was widely reported. He said the Bush administration had been "flat wrong" to reject the Kyoto accord and that cutting greenhouse gases was good for business and the planet. Within hours U.S. negotiators had returned and the conference was saved from collapse.

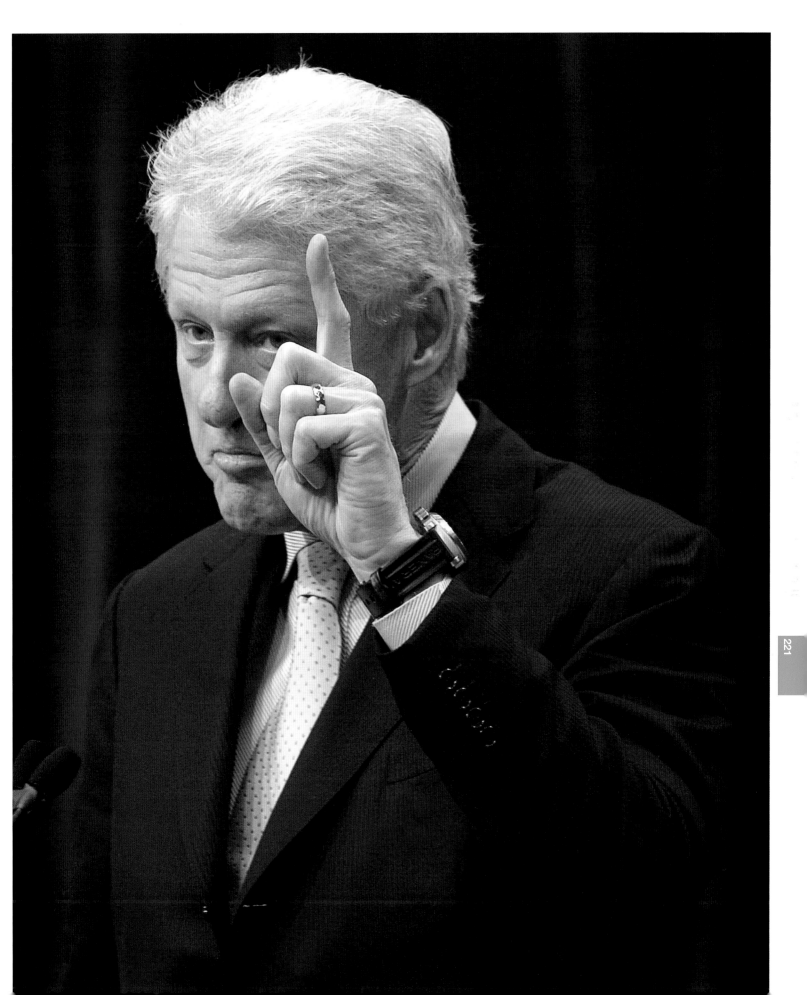

"It is not enough for us to be just caretakers of
the world that we have been given, we must
leave it a better place for future generations."

—Arnold Schwarzenegger

Above: Germany has always taken a lead in Europe on environment issues, and Chancellor Angela Merkel embraced the drive for renewable energy from the moment she took office, unleashing a massive program for energy efficiency in older housing and offices. Here the environment minister, Sigmar Gabriel (left), Chancellor Angela Merkel, and the minister of economics, Michael Glos (right), address a news conference after an energy summit in Berlin on April 3, 2006. German firms that harness renewable energy sources like the sun and wind to generate power said they planned massive investments in the coming years and were given preferential treatment.

Above: Al Gore, who appeared to the outside world to fade away after conceding defeat to George W. Bush in the battle to succeed Bill Clinton as president of the United States, re-emerged as a firebrand campaigner for climate change. The former U.S. vice president had represented the U.S. government at Kyoto and accepted a 7% cut in emissions for his country, but President Bush pulled out. Al Gore toured the world in 2006 and 2007 to alert people to what he describes as imminent disaster. Here he speaks at the 2005 Milken Institute global conference in Beverly Hills with a passion observers say he never possessed during the presidential elections.

the 2005 hurricane season did change. Many states, which were already taking action on climate change, reinforced their efforts. There was also a tide of opinion in industry that wants to seize new opportunities and begin the manufacture of new products that will cater to consumer demand for cars and electricity generation but does not harm the climate. Some sectors have expressed fears that the United States is being left behind in this new industrial boom and that the old fossil-fuel industries and car manufacturers are holding the country back.

The powerful advocates for action on climate change in both the Republican and Democrat camps were also having an effect on the White House. At the G8 summit in Germany in 2007, President Bush came under heavy pressure to move on climate. A few days before, he had proposed a new body of big polluter nations, convened by the United States, to work outside the existing U.N. structure, but this was flatly rejected by Brazil.

At the G8, under pressure from German chancellor Angela Merkel and Tony Blair in his last days of office, George Bush finally conceded he would "seriously consider" a proposal that would result in a 50% cut in carbon emissions by 2050. Mr. Blair described this as a "huge step forward," but the new French president Nicolas Sarkozy said, "We could have done better."

Although environment groups were disappointed and said the agreement was just more talk—there were no targets or timetables, just aspirations—it did mean that the United States was back inside the U.N. system. Yvo de Boer, the head of the climate-change secretariat in Bonn, welcomed President Bush's statement that he was taking climate change seriously and said it had given hope for progress at the climate talks due in Bali in December 2007. But at the G8, China and India both indicated that while they believed reductions in the greenhouse gases were vital, development was still the priority for them because they still needed to lift millions of their people out of poverty.

So many difficulties still remain. A new, certainly more positive president will not be in the White House until 2009, and climate talks have always taken years to reach conclusions. Kyoto took three years to negotiate and eight more years to come into legal force, but the clock is already ticking on Jim Hansen's 10-year timetable to take action or face planetary disaster. The battle to save the Earth clearly cannot be left to politicians alone. Something else has to happen. What might it be, and what can the rest of us do?

Above: Al Gore's message is partly that if politicians are slow to act, then the pressure must come from the public. Citizens all over the world are increasingly demanding action as they see their way of life and their future threatened by climate change. Hundreds of environmental groups and many organizations concerned about human welfare and poverty see climate change as a major threat and are putting increasing pressure on governments to take the action needed to combat it. Here is part of a mass demonstration organized at the Montreal climate talks to try and make clear to politicians that people want change.

Don't Mention Population

One of the great unmentionables in the climate debate has been population. At the 1992 Earth Summit it caused an enormous political argument behind the scenes because of the Vatican's refusal to discuss the issue of birth control.

But the fact that the population of the planet tripled in the last century to 6 billion cannot be ignored. Even with a currently falling birth rate in almost every country, the number of people on Earth is expected to rise another 2 billion by the middle of this century before stabilizing.

Predictions made in the 1960s that this amazing rise in the number of people would lead to starvation have not materialized. The population rise has been matched by what was called the green revolution, the use of fertilizers, which increased yields. There have also been large increases in the use of irrigation, and much farmland has been developed from former forests.

There are now those who think that the predictions of 40 years ago have merely been delayed by these factors, and food shortages will become normal. They give as an example the appalling drought in East Africa in 2006, which put millions at risk. The number of malnourished people in the world is increasing. It is close to 900 million in 2006.

Lester Brown of the Earth Policy Institute in Washington believes a projected increase to 9.1 billion people by 2050 is "highly unlikely, considering the deterioration in life-support systems now under way in much of the world." He points out that already 42 countries have populations that are stable or falling as a result of falling birth rates. But while extra population obviously uses more resources and puts pressure on water and wood supplies particularly, it is only part of the problem. The consumption patterns of all these people are also key. Perhaps the most telling and most often quoted example of this is the United States, which produces 25% of the world's carbon dioxide with under 5% of the population. But almost as important in the debate is diet. It takes 10 times as much water to produce beef for the North American market as it does to produce grain for the largely vegetarian diet of Indians; it also takes far more land. Cattle and other animals bred for meat produce large quantities of methane, even more than that from rice paddies. In terms of consumption and production of greenhouse-gas emissions, 100 Indians do less damage to the climate than do 10 Americans.

As with most of the problems discussed in this book, there are no easy solutions to this difficult and contentious issue. Religious beliefs and human rights issues clash, as well as the efforts to relieve poverty and preserve resources. China has tried the controversial approach of keeping population down by limiting families to one child each. Most of the rest of the world has tried educating women and improving access to health care as a way of reducing birth rates. Clean water and sanitation, which cuts the needless and appalling 5 million death toll caused by dirty water each year, mostly among poor children under five, can also have the effect of cutting family size. People no longer believe they need to have as many children to ensure that some survive. Education and health care are startlingly successful means of stabilizing population growth in some states in India (Kerala is an example). They also often help greatly in reducing poverty. This is in stark contrast to other regions in India, which have failed to help the poor in this way.

In the developed world population is already stable and in many cases falling, particularly in countries like Germany, Luxembourg, and Russia. However, carbon emissions continued to rise in all three countries in the first five years of this century despite efforts by governments to reduce them. This shows that the link between population and climate change is not as important as the lifestyle of the people involved. The rise of China as an industrial power, and the aspirations of its people to switch from bicycles to cars, is another well-quoted example. But the reality is worse. Across the world around 80% of the population will soon be in developing countries and 20% in the developed world. Those 80% of ordinary people regard development as a greater priority than avoiding climate change—that is, if they think about it as a choice at all. The International Energy Agency predicted in its 2005 report that soaring energy demand would increase carbon emissions by 52% by 2030 with current trends. Initiatives on cleaner energy could reduce demand, but even then there would be an increase of 30%. This is still too much, the agency says. The task, then, is not just how to slow down population growth and feed all those extra mouths. The second need is to develop technologies so that these new consumers, and the rest of the population, can live in the style they want without wrecking the planet in the process.

Peak Oil
and China

Previous spread: The sheer scale of everything in China, from the speed with which its economy has been growing to the spread of its already vast deserts, is hard to take in. By China's standards this is a small-scale event. These middle school pupils are congregated in a city called Guangzhou, in Guangdong Province, to attend a fair for high schools looking to recruit students before they take a high school entrance exam. More than 60,000 people took part in this event on May 15, 2004.

Above: This is perhaps how outsiders think of China: irrigated rice terraces cultivated for centuries to feed its ever growing population, now more than 1.3 billion. The effects of climate change and a scarcity of water are making rice more difficult to grow because it is a very thirsty crop.

Above: In China, regions dominated by mountains and desert are so harsh that thousands of square miles are sparsely populated. Here the Huang He or Yellow River is at the beginning of its long journey to the sea. Its waters are used for drinking, irrigation, and industry to such an extent that only a tiny fraction of its mighty flow ever reaches the sea.

Probably the best hope for the world is that the oil begins to run out. In my view the quicker the better, although others disagree, because they fear that sky-high oil prices will cause an economic slump, preventing the world from having the money to adapt to climate change.

In any event, there are signs that oil demand is beginning to exceed supply, even though some politicians continue as if it were not so. For example, China's long-term plan to 2050 includes having half of the country's projected 1.5 billion population owning their own car and being able to afford overseas travel. The plan, published in February 2006, included lifting another 80 million Chinese out of poverty and moving 500 million peasants to cities to work in factories and produce more consumer goods. This vision, which includes continuing the country's astonishing 8% to 9% annual growth rate, is impossible. If China grew at that pace, its demand for raw materials would strip the world of natural resources. Already commodity prices, for metals like copper, have increased dramatically. But the worst shortages would be in water and oil.

As far as water is concerned, China and large parts of the rest of the world already have a problem verging on a crisis. Twenty-five years before China's grand plan is supposed to come to fruition, 3 billion people in the world will be facing water shortages. At least 500 million of them will be in China, a country that is already overusing its water resources. To grow 2 lb. (1 kg) of rice uses up to 5,300 qt. (5,000 l) of water, so China is diverting its water to factories and cities to produce the wealth needed to import food rather than grow grain itself.

Even so, there is not enough water in the dams, rivers, and replenishable near-surface aquifers to go around. An estimated 100 million Chinese are already relying on non-replaceable fossil water from deep underground to water crops. It is being pumped to satisfy short-term demand. But while water is essential both to prosperity and to life, it is a local problem. If China misuses its water resources or simply runs out, it is

China that suffers most and has to solve the problem. Both Beijing and Shanghai are short of water. Shanghai is sinking as excessive use is made of groundwater. This coupled with sea-level rise, storm surges from more violent typhoons, and increased river flooding puts the whole area in jeopardy.

At the same time, the Chinese government is developing its fantastic plans to transfer water from the Yangtze River in the south to replenish the dried-up Yellow River in the north. The cost is $80 billion, and with that price tag China may be able to pipe water across country or even from neighboring states, but it cannot buy water on the world market. Water is not yet a financially valuable commodity like oil, which can be traded across the world. The Chinese, who are also aspiring to eat more meat, are relying on food imports to meet their people's wish to mimic a Western diet. Whether this is possible remains to be seen.

Oil, on the other hand, is central to the continuous and unsustainable growth in the international economy. Its continued supply and availability has been part of the built-in belief of politicians and businesses alike that the future always has the promise of increasing world trade and affluence. Part of this assumption, bar a hiccup or two, is that the oil supply is always enough to keep the price low. This assumption allows Chinese economists to believe that half of the country's population in 2050 will be driving cars and that the country's economic power will give it the cash to buy the necessary oil on the world market to power them. Even the most optimistic Americans no longer believe that. The Chinese grand plan was launched in the same month that President George W. Bush, in his State of the Union address, began a campaign to wean his

(continued on page 234)

"In little more than a decade, China has changed from a net exporter of oil into the world's second-largest importer, trailing only the United States."

—Peter Goodman, *Washington Post* staff writer, July 13, 2005

On this page: This is modern China, as the government likes to portray it: new factories and industrialization to fuel an export boom that will make the country the world's next economic superpower. Outside China it is the country's thirst for metals and particularly steel that has created a stir, not least in the rising price of scrap metal, much of which ends up providing raw material for the country's construction boom and new industries like car manufacturing. China was named the world's top steel producer in 2005 and is the world's fastest growing market for cars.

Top left: The Kwein Lin silk textile factory.

Top right: In Shanghai, a Chinese worker operates a machine for hot-rolled steel at Baosteel Group, China's largest steel mill.

Bottom left: A worker at the Anshan Steel Factory in northeastern China.

Bottom right: Some of the steel bars end up at the construction site for Wukesong Indoor Olympic Stadium in Beijing.

Top: This is the other side of China, the battle against desertification and environmental disaster. Farmer Feng Yongcun, age 74, gazes out at a 6.2-mile (10-km)-long advance column of the Gobi desert looming over his village in Longbaoshan. The sand dune was a curiosity 3.7 miles (6 km) from the cornfields of his village north of Beijing when he was a young man, but it now looks set to overwhelm his home. The "flying desert" now regularly darkens the skies of the capital Beijing, and sand rains down on the city.

Bottom: The country has always had deserts, but now they are rapidly getting larger and semiarid regions are being overtaken such as at the Jiangxi nature reserve, shown here. Staff survey the encroaching sand dunes at Poyang lake, Jiangxi Province.

Opposite: A Chinese worker washes dust off of the bushes and pavements in the center of Beijing in an attempt to clean up the city during the periodic dust storms caused by strong winds from the Gobi desert in the north. People with breathing difficulties are advised to stay indoors for their own safety during these events.

country off its dependence on oil. President Bush did not spell out exactly what this meant to the United States, but he was acknowledging for the first time that the demand for oil for the world's automobile fleet would soon outstrip supply.

Even North Americans realize that the days of cheap oil are over and that there will soon not be enough to go around. How much oil there is and how long it will last has long been debated without any firm conclusion being reached. But that is not what matters most. The important question is, and always has been, how soon the demand will exceed supply. Optimists, including the U.S. government, have always said this will be another 20 years, a figure that remains the same with the passing of the years. The so-called tipping point (as far as oil is concerned) is when prices begin to rise and never come down again because demand exceeds supply. This is always going to be years away. Pessimists, who also call themselves realists, think the tipping point may have already been reached, or will have passed before 2010. The price of oil reached a record high of $77 a barrel in 2006 and then dropped back to $52 before creeping back up to $71 at the date of going to press.

It is too early to say whether these are the first signs that the tipping point has been reached. In my view the moment is close, and at that point the price surge will continue, with some predicting that oil will soon cost more than $100 a barrel. Most important, the price will never again be low.

For the world economy many believe this will be a showstopper. Not that higher oil prices necessarily stop growth, although they will cause it to stutter, but they will turn upside down the accepted theory of the way the world works. The current economic model is that demand stimulates supply, and that natural resources

can always be found to fulfil that need. To some extent this does work with oil. Smaller marginal fields that were too expensive to exploit in the era when oil was $20 a barrel become very attractive when the price is $100. One of the resources that many believe will be utilized when the price of fossil fuels rise is tar sands. In Canada the Alberta oil sand deposits, 460 miles (740 km) north of Calgary contain an estimated 175 billion barrels of recoverable oil, second only to Saudi Arabia's 259 billion barrels, according to the Canadian Association of Petroleum Producers. Several other countries have tar sands but unlike liquid oil wells, extracting oil from sticky deposits requires a great deal of energy and it is an expensive process. This makes it a particularly difficult industry to promote in a country like Canada, which the Conservative government has already said will fail to meet its Kyoto targets. Recovering oil from tar sands not only destroys Boreal forests, which grow on the ground above, it produces toxic emissions and will considerably increase Canada's already rapidly rising greenhouse gas emissions. Utilizing tar sands would make Canada an international pariah state as far as environmentalists are concerned.

Even if the production from tar sands was ramped up it seems that the process is never likely to keep up with the continuously rising demand. The number of discoveries of new fields is not keeping pace. New oil fields, when they are found, are no longer the easy and cheap-to-pump reserves of the Middle East. Vast reservoirs of oil found just below the desert surface provided 50 years' worth of cheap oil. In the 21st century finding and then exploiting new fields involves ever more expensive deep drilling. These fields are either offshore, or in obscure and inhospitable regions, a long way from markets. Ever longer pipelines or supply

Top: President Hu Jintao of China walks on a red carpet beside a Nigerian Army officer in Abuja, Nigeria, April 26, 2006. Nigeria gave China four oil-drilling licenses in exchange for a commitment to invest $4 billion in infrastructure.

Bottom: The reason China is anxious to sign deals in Africa and elsewhere: The thirst for oil in this rapidly growing country is outstripping supply, as these motorists lining up to buy gas in Dongguan, South China's Guangdong Province on August 17, 2005, testify. Closed service stations, fuel rationing, and hour-long lines plagued China's southern manufacturing heartland of Guangdong during the year. The shortages piled pressure on the country's oil companies to boost supply.

"While many of the factors that have caused the oil price spike appear to be fleeting, there may be no respite from Chinese demand for the foreseeable future. The country's industrial base is gobbling up vast amounts of petrochemicals.... The number of cars on mainland roads—about 20 million—is expected to increase by 2.5 million this year alone."

—Matthew Forney, *Time Asia* magazine, October 18, 2004

Above: In a throwback to the car-selling methods of the 1960s and 70s in Europe and North America, Chinese models pose with cars at the opening ceremony for the new Bentley Beijing showroom on June 1, 2002. In Europe these selling methods would now attract derision or even demonstrations, but they seem to work in China, at least temporarily. Although there is a vast and growing market for cars, and increasing traffic jams in most cities, both domestic car manufacturers and importers are desperately trying to sell more.

routes make the oil expensive to deliver to customers and vulnerable to attack. Even without the possible disruption caused by the continuing instability in the oil producing regions of the world, satisfying demand is becoming a more difficult and expensive struggle. The average price per barrel can only ever be upward.

Lester Brown, the previously quoted founder and president of the Earth Policy Institute in Washington, has spent many years studying the stresses on the world's environment, particularly the effect of China's economic development on the world. In a speech to the Organization for Economic Cooperation and Development in Paris in February 2006, he summed up: "Many earlier civilizations at some point found themselves on an economic path that was environmentally unsustainable. Some understood what was happening and were able to make the needed adjustments and survive, even flourish. Others either did not understand the gravity of their situation or, if they did, could not adjust in time. They collapsed.

"Our global civilization today is also on an economic path that is environmentally unsustainable, a path that is leading us toward economic decline and collapse. Environmental scientists have been saying for some time that the global economy is being slowly undermined by the trends of environmental destruction and disruption, including shrinking forests, expanding deserts, falling water tables, eroding soils, collapsing fisheries, rising temperatures, melting ice, rising seas, and increasingly destructive storms."

Lester Brown said that despite these predictions, a large number of economists and politicians still needed to be convinced that civilization was under threat. The development of China

would change all that. For some 30 years the United States, with less than 300 million people, less than 5% of the world's population, had consumed one-third of the world's resources. That was now changing rapidly as China, with 1.3 billion people, had overtaken America in the number of people with mobile phones, television sets, and refrigerators.

China is continuing to expand consumption fast. If it reaches the same living standards as North America, as it aspires to do, then in 30 years its projected 1.45 billion people would consume the equivalent of two-thirds of the 2005 world grain harvest. China's paper consumption would also be double today's world production. China only wants one car for every two people, but one day, if it has three cars for every four people—U.S. style—it will have 1.1 billion cars. In 2006 there were 800 million cars in the whole world.

To provide the roads, highways, and parking lots to accommodate such a vast fleet, China would have to pave an area equal to the land it now has planted with rice. It would also need 99 million barrels of oil a day. Yet the world currently only produces 84 million barrels per day and is unlikely to be able to increase this by much. To make his point, Lester Brown was going beyond current Chinese aspirations for its people, but even if they became half as affluent as the average American, their consumption demands would devastate the world economy.

Yet as he pointed out, the Chinese plan is to provide for its citizens the fossil fuel–based, car-owning, throwaway economy to which most of the developing world also aspires. India, which by 2031 is projected to have a population even larger than China's, appears to be aiming in the same direction. So do the 3 billion other people in developing countries whose politicians promise them policies of expansion and consumption based on the American pattern. The problem with this model is that all countries are competing for the same oil, grain, and steel.

Fortunately, Lester Brown is not the only one who can see that this cannot go on. In fact, it has been common knowledge since 1972 when the first U.N. Conference on Human Development was held. Professor Barry Commoner, a distinguished American biologist, told the participants on the first day of the conference: "We know that natural systems which support our life cannot long withstand the wasteful destruction of the Earth's irreplaceable stores of fuel and metal consumed by factories." This was the first time that the idea of sustainable development was discussed—the idea that the Earth we leave to our children should be in at least as good a state as the one we inherited. The idea of harnessing solar power was high on the agenda as well. So was the argument from the United States that cutting back on the use of resources would cost jobs and damage living standards. Not much seems to have changed.

Yet 34 years later Lester Brown believes that he can see the beginnings of a new-style economy. He sees hope in the wind farms of western Europe, the solar rooftops of Japan, the fast-growing hybrid car fleet of the United States, the reforested mountains of South Korea, and the bicycle-friendly streets of Amsterdam. "Virtually everything we need to do to build an economy that will sustain economic progress is already being done in one or more countries," he says. He will be encouraged that in the year following his speech, many more countries have adopted new technologies—especially China—which, for example, now has solar-powered streetlights all the way from Beijing to the Great Wall, a two-hour drive away.

(continued on page 240)

237

Opposite: Although there is a growing gap between the cities and the poor in the countryside, Western-style consumerism has come to China in the form of boutiques and superstores. Here two streets, which could be anywhere in China's many megacities, are crowded with shoppers anxious to use their new-found spending power. The streets are (top) in Chengdu and (bottom) the enormous Sun Dong An Plaza on Wang Fu Jing Street in Beijing.

Following spread: While China's new urban affluence fuels a consumer boom in its rapidly growing economy, the menace of the encroaching desert is never far away. Here the legendary Forbidden City and, behind it, the Great Hall of the People are seen in Beijing on April 11, 2006, through the choking atmosphere of a dust storm. Floating dust from northern China's Inner Mongolia region frequently brings hazardous air pollution to the country's capital.

So although the problems have gotten a lot worse since 1972, at least some of the solutions have been developed and are available commercially. Just to take one example that Lester Brown mentions: the gasoline/electric hybrid cars in the U.S. The average new car sold in the United States in 2005 achieved 22 miles to the gallon, compared with 55 miles per gallon for the Toyota Prius, a dual gasoline/electric car being sold in 2007, the fist of several models now on the market. If over the next 10 years the United States replaced its car fleet with efficient gasoline/electric hybrids, oil use could easily be cut by half. This would be a huge start on Mr. Bush's aim to reduce the U.S. economy's dependence on oil.

What is clear is that the lifestyle of the Earth's existing human population is not sustainable. Take London, for example, which imports food and resources from across the world to feed its population of around 7.5 million. A calculation in 2005 showed that London uses the environmental resources of an area 120 times its own size to feed and service its population. That is an area greater than all of the productive farmland in Britain. Around 80% of the food consumed in the capital is imported, making the prices and the availability of food supplies vulnerable to an increase in oil prices.

London is not a typical city, because its inhabitants include some of the richest on the planet. But half of the world's population lives in cities, some of them three times the size of London. All of them need vast areas of agricultural land and resources to sustain them. This realization has led to a number of world cities getting together to try to avoid the consequences of climate change and over-consumption of the Earth's resources. All have similar problems, while at the same time being completely different, even when they are

geographically close together.

In Europe they include Barcelona, Berlin, Copenhagen, London, and Paris; in Asia, Beijing and Tokyo; and in North America, New York, Chicago, San Francisco, and Toronto. Mexico City, probably the largest urban sprawl in the world, and Cape Town, South Africa, are also involved—with the common knowledge that they cannot go on as they are.

Fifteen city mayors and their teams of officials met in London in 2005 to pool their knowledge and ideas on how to sustain themselves in the 21st century without destroying the planet in the process. Many of them, as has been discussed in previous chapters, are in danger from sea-level rise. Others are short of water. All use far too much fossil fuel. This last problem dominated the conference: the need to cut greenhouse-gas emissions, while at the same time transporting citizens and providing heat, cooling and light.

All of them had targets to reduce emissions, but with a remarkable variety of methods of doing so. Nicky Gavron, then deputy mayor of London, said the most rewarding part of the conference was the number of ideas and methods for improving the quality of life that could be transplanted to other cities while at the same time cutting carbon dioxide emissions. Many schemes had proved themselves in one city but had never been thought of or tried in another. London's congestion charge, which has reduced traffic by 30%, lessened air pollution, and improved public transport was one of the examples. Traffic-related carbon dioxide emissions inside the congestion zone in London had been cut by 19%, and bus usage in the city had risen 40% in five years. Congestion charging proved so successful and popular that the area where vehicles were charged $15 a day to enter

Top: Even the most die-hard of American car manufacturers realize that the days of the gas-guzzling models are over. While vehicles that use five times as much fuel as European cars are still big sellers in America, the market is shrinking rapidly and there are waiting lists for greener models. Here in Detroit, Michigan, in 2004, Toyota unveiled its hybrid gas-electric Highlander SUV at the motor show, stealing a march on the entire U.S. auto industry.

Bottom: The engine of the all-new Prius in Tokyo, April 17, 2003, gets a stand all of its own. The new Prius is equipped with Toyota's hybrid system THS II, which allows the car to run on either gasoline or electricity, as well as a system in which both the gasoline engine and the electric motor are in operation at the same time. This operation is all done automatically by the car in response to the conditions and use of the brakes and accelerator by the driver.

Above: Bicycles still outnumber any alternative means of transport in China (apart from walking), and even in the booming city of Shanghai they remain the most popular form of commuting. Try to imagine what would happen if even a small percentage of these cyclists took to cars. The entire city of 16 million people would become clogged.

was doubled in size in 2007. Exceptions to charges were given for electric cars and hybrid vehicles and proposals have been put forward to charge heavily polluting private vehicles $48 a day by 2009.

Some cities had already embarked on large-scale energy efficiency programs. New York City had already completed 164 projects with an annual energy savings of $14 million. Toronto had developed a deep lake water cooling project to air-condition large office buildings. Retrofitting 467 privately owned buildings with energy efficiency appliances and materials saved $102 million in energy costs in the city.

Beijing, already conscious of its vulnerability to water shortages, and anxious to improve its air pollution problems ahead of the 2008 Olympic Games, planned to cut coal burning in the city to less than 15.2 million tons in 2007 from 26.4 million tons in 2001. Solar, geothermal, and wind energy are being employed to provide an 80% "clean and efficient" energy mix by 2010. But during 2005 the city had an additional 230,000 cars on its roads. It countered this extra pollution by running 2,100 buses on compressed natural gas, and before the games it will have 90% of its buses and 70% of its taxis running on clean fuels.

But it was clear from a visit made in March 2007 that pollution is still very high—and unacceptable for athletes. The government's solution appears to be to stop all traffic in the city for the duration of the Olympics. This may solve the air pollution difficulty for those two weeks, but it is a face-saving measure. The traffic jams in the capital and in Shanghai and Guangzhou are already so bad the government is going to have to do far more to cure the problem or face severe economic consequences from both congestion and sickness among its population.

Left: Boats pass by Shanghai's skyscrapers in the Pudong financial district on October 10, 2001. Shanghai is China's commercial powerhouse but the city is vulnerable to climate change; the base of these vast skyscrapers is only a meter above the water level. Their basements are already at risk from a combination of river flooding, sea-level rise, and typhoon storm surges. They could be standing in water well before the end of their planned lives unless protected with new flood defenses. The city is also sinking as the water table drops.

"We see the U.N. Framework Convention on Climate Change as the only global forum for action on climate change."

—Su Wei, deputy director for treaty and law in the Chinese Ministry of Foreign Affairs, commenting on U.S. attempts to derail the Montreal climate talks, December 9, 2005

In order to try to match and exceed some of the targets of other cities, London has employed an engineer called Allan Jones to run the London Climate Change Agency. In energy circles he has become something of a legend because of his success in transforming a small English borough, Woking in Surrey, into the alternative energy capital of Europe. By using energy efficiency, solar roofs, combined heat and power, and the first fuel-cell power station in Europe, he reduced the borough's energy needs by 40% over 10 years and saved the council $9 million in energy bills in the process. The annual savings on the council's budget were $1.3 million a year.

London has the target of reducing 1990 levels of carbon dioxide by 20% by 2010, in line with the national target. However, by the same time, as a demonstration, London wants to establish one zero-carbon development in every one of its boroughs. Since 70% of the city's emissions come from buildings, retrofitting existing buildings is going to be a major part of the program. In May 2007 at a follow-up conference 16 cities including New York, Chicago, Houston, Toronto, Mexico City, London, Berlin and Tokyo concluded a deal with international banks to spend $1 billion dollars upgrading municipal buildings to reduce energy use by between 20% to 50%. Altogether 46 cities took part and pledged to reduce carbon emissions by a variety of means including introducing congestion charges for private cars similar to the one successfully pioneered in London and also planned for New York. Former U.S. President Bill Clinton said these initiatives could cut world carbon dioxide emissions by 10%.

These big city conferences are part of a much wider network of leaders of local government who are signing up to climate change targets. An organization called ICLEI, which represents local governments of all shapes and sizes, has signed up to meet and beat all national targets, wherever their members are. ICLEI, which calls itself Local Governments for Sustainability, has a membership of more than 500 councils representing 300 million people in 67 countries. At the Montreal climate talks hundreds of them signed up to a target of 30% reduction in greenhouse gases by 2020 and 80% by 2050.

Also in Montreal was Greg Nickels, the mayor of Seattle. In March 2005 he persuaded eight other mayors of U.S. cities to join him in writing to 400 colleagues across the country asking them to sign up to meet the United States' Kyoto targets, even though those targets had been repudiated by President Bush. He was astonished by the response he received from mayors of all parties. By May 2005 he had signed up 134 mayors in 34 states, all agreeing to reduce their city's carbon dioxide emissions by 7% on 1990 levels by 2012. In Montreal the number had grown again, and it reached 435 mayors, representing 61 million Americans by March 29, 2007.

This kind of commitment by local government is reflected in the surge of orders for renewable energy. While oil and coal production expanded 2% annually in the first five years of the 21st century, wind and solar energy have grown by more than 30% per year. Many mayors are finding that solar energy and small wind turbines fixed directly onto the buildings that need the power are cheaper than buying electricity from the grid.

Opposite: China has laid great emphasis on making the 2008 Olympics in the city "green." Coal-fired power stations are being shut down and renewables encouraged all over the city. Polluting public vehicles are being replaced and, as shown here, the organizing committee for the 29th Olympics is planting trees in China's capital.

Boom Time for Jobs and Sunrise Industries

Previous spread: Denmark pioneered the large-scale uptake of onshore wind energy by giving people tax breaks to install their own turbines and communities incentives to invest. That success spurred the government to look at the shallow seas around Denmark's coast to see if offshore wind farms were also viable. The wind is less gusty offshore and turbines can be bigger, thus producing more power than previously expected. This one is in the Baltic Sea near the capital, Copenhagen. As a result of the skills engineers have developed, Denmark now has a large wind-energy industry, employing 20,000 people. The country now supplies 20% of its energy from renewable sources.

Above: The Tarbela Dam on the Indus River in Pakistan was built to provide irrigation and hydroelectricity for the country and was the largest earth-filled dam in the world when it was built in 1977. The dam has been very controversial and, like many large dams, did not live up to expectations. Its life has also been shortened by rapid silting, caused by deforestation upstream, which has brought increased amounts of sediment into the dam during the annual floods. Misuse of irrigation water has also caused the salination of soil and the loss of croplands, and even desertification in some once fertile places. The loss of glaciers will severely disrupt the summer flow of the Indus.

Above: The solar panels of the world's largest roof-based solar system are seen here in the southern German town of Bürstadt when it opened on May 24, 2005. The 430,000-sq. ft. (40,000-sq m) installation produces 4.5 million kilowatt-hours per year and is one of many solar ventures in Germany, which has embraced the technology—alongside wind power. The government's aim is to get the country's greenhouse-gas emissions down to 21% below its 1990 levels by 2012 in order to meet its Kyoto target.

Developing technologies for generating electricity and replacing fossil fuels is the greatest single business opportunity of the 21st century. The race is on to refine existing and new technologies that will use the minimum amount of coal, oil, and gas, eventually replacing them altogether.

Despite the failure of many governments to give more than the minimum incentive to bring about a new industrial revolution, it is going to happen anyway. Smaller countries, like Denmark, Iceland, and most recently Sweden and Norway, are leading the way. The continued resistance and smoke screen thrown up by the fossil-fuel lobby over the last 20 years can no longer prevent progress. In the United States, California leads the way, and even George W. Bush, who has done the most to hold up change, acknowledged in 2006 that the United States needs to be weaned off its dependence on oil and a year later began to put policies in place to do so.

The good news is that the technologies to change the way the world fuels transport, keeps the lights on, cooks, and keeps cool and warm already exist. All that has been lacking has been the political will to develop and modify each system to make them competitive with fossil fuels. The right incentives would promote large-scale production and make renewables mainstream industries, wiping out the advantages of the still heavily subsidized fossil-fuel industries. Although we would all like there to be one, there is no magic bullet; there are, however, lots of different solutions. Research and development are still throwing out welcome surprises and technological advance. But what is increasingly clear is that the best renewable resources and ways to combat climate change often depend on where you live, what your needs are, and what natural resources you have around you.

Denmark, as will be discussed shortly, has made the most of wind power but is also experimenting with using numerous technologies to make one of its islands entirely fossil-fuel free, simply to see if it is possible. Iceland wants to become the world's first hydrogen economy by 2050. That means powering all of its boats and cars with hydrogen produced by electricity from surplus renewables—wind, geothermal power—and hydroelectricity.

More recently Sweden has announced a bold plan to wean itself entirely off oil in 15 years. For a cold country with 9 million people, most of whom drive cars, that is a tall order. The Swedish government simply said it will replace all fossil fuels with renewables before climate change destroys economies, and growing oil scarcity leads to huge new oil-price rises. Sweden has a head start on other countries: By 2003 26% of all of its energy already came from renewable sources, and this percentage has been rising ever since. This compares with 6% in the EU as a whole. While hydropower is the main source, Sweden also has geothermal, wind, and solar power. Forests are its main future source of local renewables. Large quantities of wood waste can be converted into biofuels.

Norway, which is the world's fifth-largest oil exporter, announced in April 2007 it planned to reduce its net carbon emissions to zero by 2050. Norwegian emissions per capita are about three times the world average, so this is a challenge. Under the Kyoto Protocol, Norway is committed to a 10% reduction by 2012, and the government already aims to cut 20% by 2020, in line with the EU, which says it will go for 30% if the rest of the world takes up the challenge. Norway plans to reduce its fossil-fuel use as much as possible with renewables, but in order to reach its carbon-neutral status, it acknowledges that the state will have to buy carbon credits from other countries—thus passing to them the financial means to invest in new technologies.

So with an increasing number of governments pledged to stimulate demand for all forms of energy-saving and new technologies, it seems

likely that business should be booming. Yet while political leaders and commentators never stop talking about globalization, it is astonishing how slowly, and how little, good and easily adaptable technologies travel. Some countries have been using cheap and easy renewable sources of power and energy for years, while 100 miles away, across a national boundary, the same resource is ignored completely. In a global economy where cheap fruit, vegetables, televisions, and textiles cross the world in ships and planes, technologies to produce renewable power or to reduce greenhouse gases have seemed unable to migrate.

That may be about to change for a rather surprising reason. The sophisticated and, to many eyes, bizarre scheme of carbon trading is channeling money across the world in order to launch new technologies. This is a difficult idea to get the brain around, involving, as it does, people buying and selling pollution. In simple terms the scheme works because some governments set limits on how much carbon dioxide companies can produce in their manufacturing processes. If companies get more efficient, or invest in advanced processes to produce less carbon, they get credits in the form of tons of carbon dioxide saved. Other companies that fail to invest or for other reasons do not reduce their emissions and so exceed their government's set targets, face a fine for every ton of carbon dioxide they emit over the limit. These companies have a possible route to avoid the fine; that is, to buy tons of carbon from a company that has a surplus, hence the carbon-trading scheme.

As part of its effort to reduce emissions under the Kyoto Protocol, the European Commission in 2005 introduced a carbon-trading scheme throughout the 25 countries in the EU. To get some idea of the size of the scheme, it covered

Above: Forests have been one of Sweden's assets, providing wood for construction and papermaking. There has always been a great deal of waste in wood production, but with modern technology all of the offshoots and smaller branches and needles of these pines can be turned into biofuels. This Scotch pine forest at Gavle, Sweden, will be managed as part of the government's plan to wean the country from its dependence on oil entirely in 15 years.

11,428 factories and other installations, allowing a maximum of 6.6 billion tons of carbon dioxide to be emitted over the three years between 2005 and 2007. The scheme got off to a shaky start because many of the EU's 25 governments set the target for their industries at a level that required only minimum energy efficiency to comply. The result was that there was no demand for surplus carbon, and prices fell sharply, giving forward-looking companies little financial incentive to exceed their targets. The targets have now been tightened, beginning in 2008, so the price of carbon futures has risen again.

Another, and rather unexpected development of this system, came about because of the Kyoto Protocol. When the protocol came into force, as has been previously mentioned, it was possible for industrialized countries to claim credits for developing clean technologies in developing countries. This was done on the basis that the atmosphere benefited wherever the carbon was saved.

European companies, some of them part of multinationals with headquarters outside Europe, realized that this scheme could be combined with carbon trading to make money. Faced with expensive investments in Europe to reduce carbon emissions to meet government targets, companies have opted instead to install cheaper clean technologies in developing countries and claim the carbon credits in Europe. China alone has more than 150 projects.

The thousands of tons of carbon saved each year in the developing world can then be turned into cash in Europe. After 2007 carbon saved under the clean development mechanism may be turned into credits in the European trading scheme and traded. For example, if a European company installed an energy-efficiency scheme in China that cost $10.00 for each ton of carbon saved, and that it was then able to sell the carbon credits in Europe at $28.00 a ton, it would be a highly profitable venture. Of course, if the price fell to $12.50 a ton, and then $8.50 as it did in 2006 when the European carbon targets were not demanding enough, then the profit would go from marginal to a loss. The EU is not likely to make that mistake again, however.

Despite these difficulties, carbon trading has touched off a remarkable change. Following years of slow progress in spreading clean technologies by other means, there were 700 projects globally in the pipeline in June 2006 under Kyoto's clean development mechanism, many of them exploiting the European and other trading schemes.

One of the pioneers has been the international lawyer James Cameron, cofounder of Climate Change Capital and adviser to the Alliance of Small Island States. He operates from the West End of London and has concentrated on a greenhouse-gas reduction program in China. This is simply because "it has huge opportunities." One of the first schemes was at the port of Guangzhou, opposite Hong Kong. An urbanization program in the region has resulted in a vast refuse problem—3,000 tons a day—rotting down to produce the powerful greenhouse gas methane. Without carbon trading the methane would have been allowed to vent into the atmosphere, adding to climate-change problems, but instead Cameron's investors paid for simple measures to capture the methane and burn it to produce energy. Since methane is 23 times as potent a greenhouse gas, ton for ton, as carbon dioxide, the value of the investment in terms of carbon saved is very large and continues year after year. This provides a valuable income in terms of carbon credits

Opposite, top: Cornell University student Maki Uchida waters the landscaped garden of a solar-powered house during the Solar Decathlon house design contest in Washington, D.C., in October 2005. The house was designed to be both functional and sustainable—part of a competition between 18 collegiate teams from the United States, Canada, and Spain. The houses were judged in 10 areas, including architecture, livability, comfort, power generation for space heating and cooling, water heating, and powering lights and appliances. Each house was also required to produce enough extra power for an electric car.

Opposite, bottom: The University of Maryland's entry in the same competition. The house was designed to resemble the path of the sun across the sky.

that can be sold on the European trading system, and at the same time the Chinese are getting paid to clean up their dumps.

"This is not a market for the faint-hearted, though, because prices go down as well as up, and it is a complex business setting up these deals and getting verification. But in two weeks of prospecting in China, we found a hundred million tons of potential emission reductions. It is a question of lining up economic interests with the public good," Cameron said.

The initial investment offered to his organization by companies wanting to gain carbon credits by investing under the clean development mechanism was $130 million, but he says that was "seed money." "If this works for these investors, there are billions waiting in the wings."

Cameron's optimism was borne out by the "remarkable growth" in the carbon market reported by Halldor Thorgeirsson, Deputy Executive Secretary of the United Nations Framework Convention on Climate Change in May 2006. He said it was a clear indication of the success of the Kyoto Protocol. In the five months since the Montreal conference, the number of registered clean development mechanism schemes had risen from 40 to 176, and there were more than 600 in the pipeline.

Richard Kinley, acting head of the U.N. Climate Change Secretariat in Bonn, stated: "We are presently fast approaching the one billion ton mark in emission reductions, which corresponds to the present annual emissions of Spain and the United Kingdom combined." The two officials said that the forward-looking businesses that were prepared to invest in these carbon markets needed stability, and they were hoping to convince national governments to help in this.

The positive and sophisticated approach by businesses to both comply with the regulations and use them to make money out of the problem is in sharp contrast to the weak political will to adopt winning green technologies on home turf. This is despite the fact that developed countries often have at least as much renewable potential as the countries being helped under the clean development mechanism. The United Kingdom is a classic example. Despite much rhetoric and often repeated claims of being a world leader, the Blair government was remarkably slow to help the renewables industry, although within it Scotland and Wales have both been more advanced. After three years of minimal support for renewables after coming into office in 1997, the government opted for boosting wind power. The saga of wind power is a good example of the difference political will can make to the development of an industry, especially given that the UK has the largest and most reliable wind resource in Europe. Perhaps because of that incentive, electricity generation using wind power was first invented in Britain. However, without government support, the industry never developed.

The Danes, living in another windy country, saw the potential 30 years ago and grabbed the technology. From the 1980s onward, the Danes created a domestic industry, and with it thousands of jobs. As the rest of the world woke up to the potential, wind turbines became the country's most valuable export. In the process the Danes became the largest producers of renewable electricity in the European Union. When the British government finally decided to encourage the technology, wind power became Britain's fastest-growing industry. Although thousands of new jobs have also been created in Britain, Denmark has provided much of the know-how and materials, further boosting its economy and

Above: Four scenes from Chongming Island, which is slightly larger than the island of Cyprus. The Chinese government aims to turn Chongming Island into a showcase of sustainable development. These traditional farmers will be protected, while Dongtan, a new city being built in the eastern part of the island, acts as a model for the future. The planners have promised to protect the wetland habitat of the island, which is fast disappearing in China. The island, in the mouth of the Yangtze River, will only have electric vehicles, while the wetlands ecosystem is protected by a special "buffer zone" separating it from the new city.

"We have the technology for electric cars, and we have other power sources. Yes, it would be expensive to make the switch-over, but we could, and we would have complete autonomy from our need for oil. We would not have to rape Alaska, and we wouldn't have to be dependent on those guys over there. In the long term, if we could free ourselves from the dependency on oil, we'd be all right."

—Brad Pitt, 2001

reinforcing the importance of early investment in newer technologies.

At the same time as boosting wind power, the British government sent all the wrong signals to other home-grown renewable industries by continuing to put millions of pounds a week into nuclear power. In 2006 this was made worse by the government's decision to join a new international debate about whether to build more nuclear stations (see next chapter). Despite this diversion, the drive for alternative energies in the UK and the rest of the world is now unstoppable. That is because the UK is running out of fossil fuels and becoming a net importer of gas and oil for the first time in a generation.

Even the UK's famously backward-looking Department of Trade and Industry has accepted that several renewable industries have huge potential. Young, bright civil servants have convinced at least some of the older nuclear power enthusiasts that there is room for several technologies. As a result, there are already many different renewable schemes contributing to the national grid. With the right research and development money, the potential is so great that fossil fuels could be cut out entirely, although so far the government has failed to grasp the opportunity and perversely stopped all grants in the spring of 2007.

It seems hard to believe this will continue given the continued political pressure to do something about climate change and the range of renewables available to the UK that are probably greater than in any other country. The UK has well developed wind- and hydro-power generation and a tiny but increasing contribution from solar.

Geothermal, which in the UK's case is mainly hot water and heat from underground, is also being exploited, and there is a small potential for more. Methane gathered from landfill sites, the technology now being transferred to China under the Kyoto Protocol, was also pioneered in the UK and is the country's cheapest form of renewable power. Garbage digesters, sewage, and manure are all used to create gas to produce energy. Straw, wood waste, sewage sludge, household garbage, chicken droppings, tires, and even condemned cattle carcases are all incinerated to generate more power. Some of these are controversial—including the burning of tires and household garbage, both of which many think should be recycled—but they all show that what used to be regarded as simply waste can be put to good use.

The main reason for the lack of development in most of these industries seems to be the old problem of insufficient financial incentives from government to bring new technology from the successful prototype to the takeoff point of mass production. The United States is another country where the dominance of the fossil-fuel industry and the government's continued subsidies for oil and coal in the form of tax breaks mean that cleaner renewable industries have struggled. That, too, is beginning to change.

A recent example is concentrated solar power (CSP). In essence, this is a simple technology. Fields of mirrors are used to collect and concentrate the sun's rays to create heat—usually concentrating it on a liquid, most often water, to raise steam, which drives turbines and generators to create electricity as in a conventional power station. This is one of the most interesting, exciting, and most promising technologies as far as meeting the challenge of climate change is concerned. Again, like so much in the renewables field, it is not new; however, when the price of oil is above $50 a barrel, this technology becomes economically competitive with fossil fuels. The U.S. Department

Left: This is one of only four pump storage facilities in Britain. The 440-megawatt Cruachan power station on the banks of Loch Awe in Argyll, Scotland, looks like most hydroelectric schemes producing electricity as the water stored behind the dam runs through the turbines. At slack times, however, when there is no demand for electricity, the surplus power generated is used to pump the water back up to the dam from Loch Awe. This is a way of making sure there is always plenty of water available to provide power at times of maximum electricity demand. Until scientists think of another way to store electricity, apart from low-voltage batteries, Scottish Power, the owner, believes that pump storage is the most efficient method of harnessing power from renewables so that it can be used when most needed.

of Energy developed various types years ago that were tested in California's Mojave Desert. Surprisingly, the government never exploited the technology to wean the country from its coal dependency. In the summer of 2007, a new CSP plant opened in Boulder City, Nevada, providing 64 megawatts of power—enough to supply 15,000 homes annually, or the same savings in carbon dioxide as removing 17,000 cars off the road.

But that effort had already been eclipsed by a "power tower" in the center of a field of 624 mirrors near the southern Spanish town of Sanlucar. There the plan was to extend and develop the power project to provide 300 megawatts of electricity—four times the size of the U.S. installation. This is still small compared with the output of a large coal-fired station, but these schemes are also tiny compared with the potential for this underutilized renewable. If only 1% of the world's deserts were covered in this kind of mirror technology, it would provide enough energy to satisfy the world's current electricity demands.

Like all of the other technologies explored in this book, there is not just one that suits every climate and every circumstance. However, for desert and sunny areas where marginal land is not used, CSP will make a huge contribution to reducing greenhouse-gas emissions. Apart from California, the places that could obviously benefit are large swathes of the southern United States; China, particularly Beijing, which is close to the Gobi desert; and all of Australia. The Middle East deserts are also perfect for this technology and could use surplus power to desalinate water.

Perhaps for Africa it is both a local solution to the lack of energy supply and an export opportunity. The World Bank is funding projects, and a consortium is already looking at building CSP plants in empty desert regions of North Africa and laying cables across the Mediterranean to southern Europe to supply power. The Trans-Mediterranean Renewable Energy Cooperation (TREC) is exploring the idea of desalinating seawater with surplus energy and is using the shade of the giant mirrors to generate thermal cooling. Various technologies for storing power for nighttime use, including the creation of hydrogen for use as a fuel for turbines, are also being considered. The German Federal Ministry for Environment has commissioned reports into how this technology could be exploited to cut Europe's carbon emissions by 70% by 2050, phasing out nuclear power at the same time.

In Britain, where the government has said it is in favor of renewables but fails to support them, most observers expect to see the technologies developed elsewhere and then reintroduced into Britain. Among those could be wave power, and one of the new technologies with great potential, tidal turbines. Wave power has been in development for 30 years, and a bewildering variety of machines has been built, but many of the most promising prototypes have not succeeded. The greatest problem has been how to deal with storms and the unpredictable destructive forces they generate.

The installation of the first commercial wave-power generating station off the coast of Portugal in 2006, using British technology, will almost certainly be the start of a revolution as important as the wind industry. This station relies on the movement of a series of devices floating on the surface of the sea to harness the power to drive turbines. The UK is now belatedly testing four different types of wave-power machines in the Bristol Channel, hoping to select a winner that will provide power along the parts of the west coast that are exposed to the Atlantic swell.

Opposite and above: Vehicle congestion and pollution from the exhausts of cars, trucks, taxis, and buses are both economically damaging and the cause of health issues and the premature deaths of thousands of people each year. After years of inaction, local governments and city mayors are at last taking action to protect people who live in their cities. Ken Livingstone, London's mayor, took the bold step of introducing a charge for private cars entering the city center. The congestion charge cut traffic, reduced pollution, and allowed public transportation to move more freely, encouraging far more people to use public transportation. Other cities in Britain are considering similar action. Many cities now use buses running on natural gas. London buses have also switched to cleaner fuels. As a result, air pollution has fallen dramatically.

One of the least publicized but least problematic and the most certain winner of the newer renewable generators is the use of undersea turbines. Some work in exactly the same way as wind turbines, using tidal currents instead of wind. Others use underwater sails to capture the force of the tide to provide the power to turn the turbines. There are two main advantages of these undersea technologies. Unlike wind, tides are entirely predictable months and years in advance, so the output from the turbines is entirely reliable. The second is that the turbine blades, or sails, need to be only one-fifth of the size of those of wind machines to generate the same amount of energy. That is simply because water is far more dense. Tidal machines are now being developed along the Scandinavian and UK coastlines with a variety of consortiums interested in the potential. Placed on the sea bed out of sight, they face none of the difficulties the wind industry has from people who believe that turbines ruin the landscape.

So, as with wind and waves, Britain is lucky to have such a large potential for another winning technology. The curious fact is that the UK's Department of Trade and Industry appears to be doing its best to play down this technology, officially claiming it could only provide 4% of the country's electricity. This is a ridiculously low estimate according to scientists and businessmen who have been studying undersea currents. They say that with so many offshore islands, most surrounded by tidal races entirely suitable for undersea turbines, this renewable resource could make a large contribution to Britain's energy needs well before nuclear stations could be built.

Among the places already surveyed are the Channel Islands, which have an excess of the resources they need for their own electricity production. This raises the possibility of using surplus electricity for other purposes or for export. It is only a short distance via cable to supply mainland France. The Bristol Channel, which has some of the highest and strongest tides in the world, could supply both south Wales and the English west country. The only missing ingredient to revolutionize Britain's power industry and make the country a net exporter of power is the political will to make it happen. Why this is so is a mystery, although the mood could change, as it did with wind power, once the technology is developed. As the Danes have discovered, making the machinery for export can provide thousands of new jobs.

Perhaps the strangest aspect of the climate-change debate over the past 20 years is that many of the easiest options for cutting fossil-fuel use, which save money and reduce greenhouse-gas emissions, have been around for decades but have remained seriously neglected. Everyone mentions energy efficiency as the first priority, but few countries have taken it seriously. Using power efficiently is such an obvious money saver and one area where industry has quietly gone ahead without needing any political lead. British Telecom, for example, realized it could save money by cutting energy use as far back as the mid-1980s. Once it had devoted staff and time to considering the problem, the company found it far more rewarding than expected. Ahead of its time the company set a target of reducing carbon emissions by 25% by 2010 and purchasing 10% of its electricity from renewables. By 2004 combined reductions of emissions from energy use and transportation had reached 46%. The enthusiasm with which staff became involved in the drive to switch to renewable power and save greenhouse gases improved staff morale and encouraged the company to attempt to supply all of its power needs from renewable resources.

YOU CONTROL CLIMATE CHANGE.

www.climatechange.eu.com

TURN DOWN. SWITCH OFF. RECYCLE. WALK. CHANGE

In the next 20 years, the world will grow by nearly one and a half billion people.

So how do we feed their appetite for energy?

will you join us.com

Chevron

Human energy™

Other examples include HSBC Bank and British Petroleum (BP), a company that has spent millions of dollars projecting a new image of itself as Beyond Petroleum. Although BP announced record profits from record oil prices in 2006 and is still regarded with suspicion by environmentalists because of its recent safety record and some of its projects in Alaska and the Caspian region, its television advertising alone has done much to raise the profile of climate change in the UK. This change in a company that until the mid-1990s was a major contributor to the Climate Change Coalition, the pressure group campaigning against attempts to reduce coal and oil use, shows what can be achieved. Since 2001 the reduction in emissions from the company's energy-efficiency projects has reduced emissions by 4 million tons, the quantity of pollution generated by a European city of half a million people.

Subsequently, the company has decided to invest $370 million in energy-efficiency measures and expects to get more than that back from energy savings. The company already has a solar business, and as part of its "beyond petroleum" image, the company has pledged to extend its global wind-power capacity to 3,000 megawatts by 2015. The company says this is enough to power 1.5 million typical homes in Europe.

Compared with Exxon Mobil, the world's largest oil company, BP, has made great strides in the right direction. There can be no doubt that any or all of these oil companies could invest vast sums in renewables and hardly notice it. Exxon surpassed BP, Amoco, and Shell by declaring the largest profit ever recorded in the world in 2006. BP and its rival Shell have made strategic decisions to invest in the future, while Exxon Mobil has made large energy-efficiency savings but turned its back on anything but fossil fuels and cancelled its renewables program.

Above: Advertising is a powerful medium. The European Commission believes that by bypassing governments that are not doing enough to combat climate change and by appealing directly to the public, it can change behavior. It is amazing how much carbon dioxide could be saved if everyone did what the poster at the top suggests. The Chevron poster at the bottom directs the public to its website and shows that for the first time an oil company is prepared to admit that one thing is clear: The era of easy oil is over. The company says that many of the world's largest oil fields are "maturing"—in other words, coming to the end of their maximum production, and that new finds are harder to exploit. The company set up their special website—which has been a great success—to engage the public in a continuing debate about what to do next.

"Wall Street is waking up to climate change risks and opportunities. Considerably more of the world's largest corporations are getting a handle on what climate change means for their business and what they need to do to capture opportunities and to mitigate risks. This all points to a continued elevation of climate change as a critical shareholder value issue for investors."

—James Cameron, chairman of the Carbon Disclosure Project, 2006

BP has also decided to concentrate much more heavily on gas, a far cleaner fossil fuel than oil and coal, but the company also makes it clear that gas is only an interim measure toward a larger renewables portfolio. There is no doubt that the personal convictions of Lord Browne, the group chief executive, who was then at the helm of BP Amoco, drove the climate change program. It can only have helped his cause with the rest of the board that many analysts believe that the era of cheap oil is over forever and that dwindling supplies mean that the tipping point where demand exceeds supply is close. This alone would make sense of the company's decision to invest heavily in solar power. BP is now a major player in a market that is growing as fast as it and other larger manufacturers can make and deliver the solar panels. For now, though, the company, in its duty to shareholders, will continue to defend its interests in oil as long as fossil fuel remains the bulk of its business and the mainstay of the company's profits.

But another positive sign is that one of the schemes the company is pioneering is carbon storage. This involves capturing the exhaust gases from burning fossil fuels, typically from coal-fired power stations, and pumping them under pressure down old oil wells, coal mines, and saline aquifers. The idea is to store carbon dioxide permanently underground. One of the target areas for this is older oil wells that are nearing exhaustion, a technique already used in depleted oil reservoirs in Texas. As a bonus for BP, the introduction of carbon dioxide under pressure in the North Sea will lengthen the life of the fields by pushing more oil out of the ground. The profit from this will help to pay for the technology to capture the gas and pump it out to sea.

Experiments are taking place in the United States and elsewhere to pump carbon dioxide into seams of coal that are too geologically unsafe to mine. Clinging to the coal are large

Above: One hope for staving off the worst effects of global warming while we await alternatives to fossil fuels is carbon sequestration. This giant apparatus in the Algerian desert is designed to inject 1 million tons of captured carbon dioxide gas into natural gas wells. Once the natural gas is pumped out, the carbon dioxide will replace it, and the well will eventually be sealed. It is hoped the carbon dioxide will be stored for hundreds, possibly thousands of years, while mankind attempts to get the atmosphere back into balance. British Petroleum (BP) is among those investing in this technology, which would help the company avoid carbon taxes and notch up carbon credits to trade internationally.

quantities of methane, but carbon dioxide fixes to coal more easily than methane, and the theory is that methane will be released to be burned as a fuel. Again the methane gained from the process would pay for the cost of the carbon sequestration. This is likely to be controversial because it is hard to calculate exactly how much the atmosphere gains from pumping carbon dioxide underground only to release methane to be burned. If the industry is hoping to make money by selling carbon credits, they will need a robust scientific case.

A third method of carbon sequestration is to pump carbon dioxide into saline aquifers, which are plentiful across the planet. This is already being done by the Norwegians who must strip carbon dioxide out of their natural gas pipelines to keep it from turning into dry ice when being pumped ashore under pressure. The operation costs only a tiny fraction of the total cost of production and prevents the carbon dioxide equivalent to the output of a small coal-fired power station from reaching the atmosphere.

Politicians are very interested in carbon sequestration, which is seen as a painless way of solving the climate problem without having to stop burning fossil fuels; however, there are problems with the technology. First, it is expensive; second, it is not yet certain that some of the carbon dioxide does not leak and find its way back to the atmosphere; and third, the number of suitable oil wells and even coal seams that could be used to recover energy and part of the cost are limited. In other words carbon sequestration is not the magic bullet.

On the other hand, energy efficiency is something everyone knows will work, but most do little about it. However, the performance of the multinational companies described above shows that whatever business you are in, irrespective of where you operate in the world, energy efficiency can save hundreds of millions of dollars through simple measures.

The other battle is to make the best use of energy produced by fossil fuels. This involves the widespread adoption of combined heat and power. That is a simple system where the surplus heat from generating electricity is used for district heating or other useful purposes rather than allowing it to be wasted by letting it escape into the atmosphere. Combined heat and power plants still work well on a large scale in former eastern bloc countries. They are even more efficient with small generators in factories, groups of office buildings, or housing estates elsewhere. In some so-called advanced countries, notably Britain and the United States, this is a badly underutilized technology. It is among the many examples across the world where some straightforward methods of saving fuel and carbon emissions have been embraced in one country and ignored in others, when they could be universally effective.

One of the countries that has consistently made the most effort to be more energy efficient is Japan. They began looking at alternatives to fossil fuels half a century ago. This was driven by Japan's vulnerability to oil shortages; it has no reserves of its own. Initially government and industry concentrated on energy efficiency—such devices as low-energy lightbulbs and combined heat and power. Ever since the 1973 oil shock, Japan has been the most energy-efficient large country in the industrialized world. In the last 10 years, in a bid to meet its demanding 6% Kyoto reduction targets for greenhouse-gas emissions, the country has made another series of new strides.

Despite its energy-efficiency reputation, Japan, along with the rest of the world, still acknowledges that it wastes far too much energy because of

Above: The magnetic levitation train developed by Central Japan Railway Company. The train is held above the track by the opposite forces of giant magnets and also propelled along by a linear magnetic motor. It is capable of remarkable speeds because there is no friction. Despite an 11.4-mile (18.4-km) test track in Tsuru, west of Tokyo, a $2 billion investment, and 40 years of development, the railway company has yet to persuade the government to build a commercial line between Tokyo and the western city of Osaka.

inefficient lighting, heating, and cooling. With the glass tower blocks that dominate central Tokyo, cooling is now far more important than heating. Since the Kyoto agreement was signed, the largest combined heating and cooling system in Japan has begun operating at Harumi Island, in Tokyo. Triton Square is a building complex housing thousands of office workers facing Tokyo Bay. The system is based on a water tank the size of 50 swimming pools. Heat exchangers provide hot and cold water as required, using the massive capacity of the tank to store unused energy in the water. Electricity is used only at night, when tariffs are lower, to replenish the system if necessary, further cutting energy costs. It has cut carbon dioxide emissions by 60% and increased energy efficiency by 40%.

While Europe has led on wind power, the Japanese have been concentrating on solar. The huge Sharp company has long been the world's largest producer, installing its first solar-powered lighthouse in Nagasaki in 1963; now there are 589. Solar electricity costs have tumbled to one-third of their price in 1993 and are still dropping as the technology improves and mass production reduces the unit price. Exports to Germany are keeping Sharp's factories at maximum production, although Germany has now developed an industry of its own.

Japan, along with other countries with volcanic regions, has also long exploited geothermal power. It is not a new technology—Italy pioneered producing electricity from geothermal power in 1905. Hot rocks near the surface, stable enough to allow water pipes to be installed, are exploited to produce steam to drive turbines. The same technology is also used to produce hot water for district heating schemes. In Iceland 93% of homes are heated with geothermal hot water and surprisingly France, more famous for its nuclear power stations, has 70 geothermal

plants producing both central heating and hot water for 200,000 homes. The same techniques can be used to heat greenhouses to grow food in the winter and for fish farms for species that grow faster in warmer water.

But while the idea of geothermal technology is simple, it requires expertise to exploit, and much of the heat that could be harnessed is wasted, partly because it seems so plentiful and costs nothing. Realizing that far more energy could be utilized, the Japanese have developed turbines that can reuse steam at lower temperature and pressure. This produces far more power with the same quantity of heat. It is estimated that Japan's geothermal potential is 69,000 megawatts, enough to produce one-third of its electricity needs. Japan uses the water for thousands of spas, swimming pools, and to heat hotels, but it is still a badly underutilized resource.

At the moment the United States with 2,000 megawatts and the Philippines with 1,900 megawatts lead the world in geothermal energy for electricity. Altogether 25 countries use it, and the industry is expanding all the time. All around the Pacific Rim and in many countries where hot rocks are close to the surface, there is great potential. Among the countries with an obvious resource are Chile, Peru, Ecuador, Colombia, all of the central American countries, Canada, Russia, China, South Korea, Indonesia, Australia, and New Zealand. A U.N. report has also found potential in the Rift Valley area of Africa and a pilot plant is now being built there. In the eastern Mediterranean there are also unexploited hot rocks.

Without the know-how geothermal is often a difficult technology to exploit, so there is further opportunity here under the Kyoto clean development mechanism, coupled with carbon-trading schemes, to transfer this technology to less developed nations, thereby saving large quantities

Top: Japan has created and maintained a steady lead in the solar revolution, with its prime minister leading by example. In this picture taken in April 2005, the then Japanese prime minister, Junichiro Koizumi, center, makes toasts with his predecessors Yasuhiro Nakasone, left, and Kiichi Miyazawa during a luncheon held to unveil his mansion. After three years of renovation work, the mansion features solar-power panels that cover the roof of the 8,400-sq. yard (7,000-sq m) four-story building.

Bottom: The solar calculator is to be followed by the solar mobile telephone. This prototype, displayed in the Wireless Japan 2005 exhibition in Tokyo, can be charged without needing to be plugged in. It is one of a number of solar phones that are expected to be on the market soon.

of carbon dioxide. To give one example, Indonesia has more than 222 million people and 500 volcanoes—more than enough hot rocks to provide hot water and power for the whole country. A grand plan to build 11 geothermal power stations exists on paper, but the country lacks the resources to exploit this use of carbon-free technology.

As with the potential use of wave and undersea turbines in Europe, Japan is also looking at the oceans. It is in the seas of the tropical regions where the other great promise for alternative technology currently lies. Japan has one technology, "ocean thermal-energy conversion," which also has enormous potential, and not just to produce electricity. It can also produce fresh water, hydrogen, and lithium, a valuable industrial material.

Essentially the idea is simple: Warm water on the ocean surface is hot enough to turn liquid ammonia into a gas, powering turbines as it expands. Cool water pumped from the deep ocean is then used to turn the gas back into liquid ammonia. Using a series of heat exchangers and pumps, the system can also be used to produce drinking water and extract hydrogen from seawater. The deep ocean water contains lithium and other useful substances that can be extracted, and it is also high in nutrients—something that has implications for fisheries, including the potential development of large-scale ocean fish farms. The Japanese Fisheries Agency is cooperating with Saga University, the developers of the system, to see if the discharge water from the energy-conversion unit can be used to stimulate new fisheries.

The power available from this source is huge, especially in the tropics where water temperatures are the highest, and particularly in small-island Pacific states and the Caribbean, where the cost of fossil-fuel imports is large, and fisheries are important. Hydrogen to power fuel cells is seen by many as the front-runner to replace fossil fuels for transportation. For many poor coastal countries and islands, this could be the answer. The only "waste" from burning hydrogen is pure water. Meanwhile, Japan has been working on the most difficult area of transportation, too. As we have already stated, the Toyota Prius is already well known internationally as a hybrid car that uses its braking system to generate electric power. The new improved model hardly uses gasoline at all in some driving conditions and claims a 43% reduction in carbon emissions even over the previous model. The Japanese have also harnessed this technology to help drive their trains. Tokyo Railway's newest train model consumes 40% less energy as a result; quite a savings since the company carries 2.7 million people a day.

Transportation, because of the need for alternative fuels to power vehicles, ships, and aircraft, remains one of the most difficult areas. The surge in oil prices has begun to concentrate minds, not just at government and industry levels but also among consumers. The financial troubles of some of the world's great carmakers like Ford and General Motors are due as much to building the wrong kind of gas-guzzling models as they are to the high wages and pension schemes of the workforce. As with electricity production, the race has been on to find alternatives to gasoline and diesel fuel. As so often happens in a world dominated by big business, the most promising interim solution comes from a developing world, in this case Brazil.

Brazil is the world's largest producer of ethanol, an industrial alcohol based on waste products from the sugar industry. Now more than half of the cars sold in Brazil can use ethanol as fuel. The country burns 4 billion gallons annually.

Left: China has relocated at least 1.3 million people from the areas along the Yangtze River affected by the mammoth Three Gorges Dam project, which is intended to tame the river and generate vast quantities of electricity. This is probably the most controversial big dam project of all time, with many doubting that it will ever live up to Chinese government expectations, partly because of the river's silt problems. The loss of so much land and river wildlife is also lamented. Chinese officials maintain that the dam is an alternative to building many coal-fired power stations, which would otherwise make climate change far worse. The dam came into operation in June 2006.

Four in ten vehicles run entirely on ethanol, and the remainder use blends of up to 24% ethanol and 76% gasoline. The country has taken 30 years to develop the technology, which began as a result of the 1970s gasoline price rise. Exports of both the fuel and the plant to make it are a potentially big moneymaker for Brazil as well as providing a renewable method of running vehicles and avoiding oil imports.

A variety of fuels produced from plants, trees, wheat, corn, sugarcane, and oil seed rape, producing everything from alcohol to cooking oil, is now being used to power vehicles. It is perhaps ironic that in 1898, when Rudolf Diesel first demonstrated his compressor ignition engine at the World's Exhibition in Paris, he used peanut oil to power it, thus making it the world's first biodiesel. Vegetable oils were used in diesel engines up until the 1920s, when engines were altered to enable them to use a residue of petroleum. At that point biodiesel was all but forgotten.

Ethanol, which is principally alcohol produced from sugarcane, can also come from a variety of grains, including wheat and corn. Biofuels are being introduced across the world from different local crops. In Europe and the United States they are mostly used as a 5% to 10% supplement to existing diesel or gasoline, but in many countries some vehicles run entirely on nonfossil fuels. European legislation has driven a market for biofuels by insisting that they take up 5.57% of all transportation fuels by 2010. But because Brazil is so far ahead of the game, it is beginning to cash in on its technological lead with an export business to China, which wants to reduce its dependence on oil, too.

One of the big problems of this biofuel revolution is that many of the crops being used are also food crops. Already the United States's demand for corn for fuel production has been blamed for the rise in the price of the Mexican staple food, the tortilla. This has also been one of the factors in the rise of the price of wheat, along with a prolonged drought in Australia. In Europe biofuels are being directly blamed for the destruction of the Indonesian and Malaysian rain forest and the resultant loss of species like the orangutan. The European Union has set targets for 10% of all vehicles to be run on biofuels, but much of that oil comes from palm oil plantations carved out of pristine jungle. So without proper management the move to biofuels could be a potentially disastrous incentive to cut down even more forests for farmland and plantations.

A better alternative would be to find crops that can be turned into biofuels that are waste from food crops or not food crops at all that can grow on land that would otherwise not be productive. Inedible vegetable oils, mostly produced by seed-bearing trees and shrubs, provide an alternative. One example is *Jatropha curcas,* a tree that grows in tropical and subtropical climates across the developing world. It happily grows on non-agricultural and marginal land and produces seeds that can be crushed to produce vegetable oil. It can be used instead of mineral diesel or mixed with it, producing a much cleaner-burning fuel. This is a tree that already grows widely in Africa, India, Southeast Asia, and China, often in arid lands, and requires little water or nutrients. The tree matures after only three years and lives for 30, providing the potential for early returns. One company, D1 Oils, which floated on the Alternative Investment Market of the London Stock Exchange in 2004, has already gone into production.

The advantage for countries, regions, and individual farms that invest in biofuels on otherwise unproductive land is that equipment for producing biodiesel is both small scale and portable. It is therefore possible to hire a

Opposite: Workers load palm oil fruits onto a truck at a plantation near Kuala Lumpur in Malaysia. Prices for Malaysian crude palm oil continued to rise in 2006, since palm oil is used in the production of biodiesel, one of the world's fastest-growing industries. Although biodiesel cuts the need for fossil fuels and reduces carbon emissions, it does have a serious downside because more tropical rain forests are being cut down to create new palm oil plantations.

On this page: In Brazil it is the sugarcane that is being converted into another alternative to fossil fuels—ethanol—in which the country has a world lead. Ethanol is one of the cheapest and most dependable fuels derived from renewable sources. Three-quarters of the cars now being produced in Brazil can run on either ethanol or gasoline, or any mixture of the two. At the top a plant producing new fuel. At the bottom workers taking a break after harvesting a field of sugarcane at the Sao Tome plantation in the southern Brazilian state of Parana.

biodiesel plant to arrive to coincide with your harvest, turning crops into a cash product on the spot. The technology could make some large-scale oil refineries unnecessary.

This typifies one of the most significant battles, both in electricity generation and other technologies. It is the struggle between large centrally produced power and local solutions. Governments and companies have always been in love with the grandiose. Giant hydroelectric dams and vast concrete nuclear stations go along with highways, bridges, and sports stadiums as proof of political potency.

The new generation of technologies, which offer the best hope for the future, are the opposite. Some of these are discussed in the final chapter and involve some of the options for those individuals who want to save the planet. This does not mean that big business and large-scale manufacturing are not needed. The idea is to produce electricity as close as possible to the point where it is used. At present, in most of the world, power plants are so far away from the customer that 10% to 15% of the power is lost in transmission, and all the heat generated is vented to the atmosphere. One of the major advantages of solar panels, fuel cells, small-scale hydroelectric power, mini wind turbines, and combined heat and power plants for individual homes is that mass production brings the unit price down for each customer to install it.

The second advantage of these systems over the large scale is that the distance between the generation of power and the point of use is meters not miles. Solar panels and combined heat and power plants do not need a grid, and all of the power produced is used either directly or by nearby neighbors. In all grid systems, no matter how good the cabling, some power is lost. The other possibility with renewables, which by definition do not rely on fossil fuels, is that the raw material—energy from the sun, wind, waves, or heat from the ground or oceans—is free. That often means that even when the power is not directly needed for electricity to keep the lights on or for heating, power is still being produced.

Since electricity cannot readily be stored except in batteries, this could be a disadvantage. It also raises all sorts of possibilities and potential advantages, although application requires ingenuity. Perhaps the simplest example of the use of spare power is for family cars. Electricity can be used directly for an electric car, scooter, or bike.

Many believe the biggest potential is to produce hydrogen by electrolysis from wind farms or from solar systems when power is not needed directly. Obviously, producing hydrogen in this way is an expensive and not very sensible use of electric power if it is produced from fossil fuels. If, on the other hand, hydrogen were automatically produced when electricity from wind power, solar, or wave power was surplus to requirements, then hydrogen would be able to replace fossil fuels in any number of applications. Hydrogen fuel cells for cars are in development. Fuel cells can also be used to produce electricity. With pure water as their by-product, fuel cells would seem to be an ideal technology for desert kingdoms, which are currently reliant on oil. Solar panels or concentrated solar power could produce electricity, with the surplus power used for the production of hydrogen, which would power the turbines at night, producing water as a by-product.

There are many possibilities to harness these readily available technologies. They require only inventive engineering skills to understand the possibilities and make them work. But to bring the unit price down so individuals and local communities can take advantage of the technology, politicians and companies need to act together to ensure large-scale mass production.

Above: On many remote Pacific islands electricity has not been available unless islanders buy expensive and noisy generators. Then they have to import and beg for regular supplies of fossil fuels, which may be beyond their means. They can, however, leapfrog the technology of the 20th century with photovoltaic panels, which turn their plentiful sunshine into electricity without breaking into the natural sounds of the waves breaking onto the beach. This solar panel on Tobi Island, in the Pacific's Belau Islands, is one of several that produce all of the electricity used on the island.

Opposite: Solar cooking is catching on in many parts of the world, particularly in the tropics where the sun is strong. Here students in Brazil test the heat given off by one of the demonstration stoves provided by a Greenpeace Positive Energy Tour in 2004. The basic idea is to use shiny or polished steel as a mirror to concentrate the sun's rays onto a cooking pot. The pot boils almost as quickly as it would over a fire, and it is possible to cook a complete meal without using any fuel or other energy source apart from direct sunlight. Its use also saves families from traveling miles on foot in search of wood for fuel.

271

Voodoo Economics

Previous spread: Bulgarian schoolchildren put on their gas masks during an exercise in the town of Kozloduy near Bulgaria's nuclear power plant, some 125 miles (200 km) north of the capital of Sofia on October 9, 2002. What use they would be in the event of an accident at the Soviet-made nuclear power plant is not clear, but the safety of the plant is a cause for concern. The future of the plant, which accounts for up to half of the country's electricity output, is a sticking point in talks with the European Union. The EU would like the plant closed, while some Bulgarian politicians are making it a matter of national pride to insist on building a second one.

Above: Perhaps the most controversial nuclear plant in Europe is the sprawling Sellafield site on Britain's Cumbrian coast, seen from the local village graveyard at Seascale. The site is the home of two nuclear reprocessing plants, which produce plutonium and depleted uranium from spent nuclear fuel. Although a very small quantity is reused for a nuclear fuel called MOX (mixed oxides of plutonium and uranium), the rest is stored on site under armed guard. The site was the scene of a nuclear fire in 1957, which left behind a contaminated nuclear reactor that is still too dangerous to demolish. Large quantities of nuclear waste are stored on site.

Above: France has the greatest reliance of any country on nuclear power and uses its major rivers for cooling the water for some of its many stations. Bugey, on the banks of the Rhône River in Saint-Vulbas, near Lyon, is one of many inland that are vulnerable to being shut down during droughts and warm weather. The plant warms the Rhône so much that aquatic life in the river is threatened. During the 2003 heat wave, state-owned Electricité de France cut its power capacity by 4,000 megawatts as a result. This was despite the fact that the government had previously relaxed environmental rules allowing nuclear power plants to put the water they use back into rivers at higher temperatures than usually permitted.

Building large numbers of new nuclear power plants is seen by many as the solution to global warming. Indeed, the industry has long pushed the argument itself. It is, they say, the only large-scale power-producing technology that does not emit carbon dioxide.

This is one of the most contentious issues surrounding climate change. Many countries have rejected the technology entirely, while at the other extreme France has built so many nuclear stations that 79% of its electricity comes from that single source. The problem of what to do with the waste, the potential for developing nuclear weapons from the same technology (which has caused so much trouble with Iran), and the nuclear stations' vulnerability to terrorism all add to the sharp debate.

Some scientists, including the UK's chief scientific adviser Sir David King, have urged returning to nuclear power as a solution. He believes that the UK cannot meet its carbon dioxide reduction targets without building new nuclear plants. Partly as a result, Tony Blair, then prime minister, ordered a new energy review in Britain in 2006. Political commentators claimed the only object was to overturn a study less than four years old that concluded nuclear power was not the answer. The new study duly concluded that a new generation of nuclear stations was needed, but this decision was subsequently overturned by the UK High Court. Judges concluded that the decision was not the result of a genuine consultation. Only time will tell whether this enthusiasm for new nuclear plants will survive.

The industry itself, on both sides of the Atlantic, has long been in the doldrums. Since the Three Mile Island incident in Pennsylvania in 1979 and the catastrophic Chernobyl disaster in the Ukraine in 1986, there has been public fear of nuclear power. Apart from voter opposition, the world's big financial institutions have been unwilling to back the building of new nuclear plants. Without massive state subsidies for disposing of waste and providing insurance, the industry cannot compete.

James Lovelock, inventor of the Gaia Theory, thinks the danger from man-made climate change is so great that building new nuclear plants to solve the problem is the lesser of two evils. The main environmental argument against this is that a plant may last 40 years, but the radioactive waste produced remains dangerous for many thousands of years. So far, despite 50 years of trying, no one has come up with a solution to store or dispose of the waste safely. A possible exception is in Scandinavia, where they use purpose-built deep-rock caverns. Everywhere else, waste is stored in a variety of concrete bunkers, all of them with a much shorter life than the waste itself. This radioactive waste will remain dangerous long after the concrete has disintegrated. There is a strong chance that climatic conditions will have changed so much within 300 years that the remaining radioactivity will be released, with unknown effects on future generations.

But the argument that will, or should, kill the large-scale nuclear plan is that it is not economic. There are lots of other better and cheaper solutions. That does not mean that the economic argument is an easy one to make. The nuclear industry still suffers from what its detractors call "voodoo economics." The sums put forward by the industry certainly do not tell the whole story, and they do not add up. What is astonishing is that spurious figures are frequently believed by governments and journalists alike. At present the industry claims it is almost competitive with coal, gas, and wind power, and if enough stations were built, economies of scale would bring the price down further. Better still, if the fossil-fuel generators were made to pay a carbon tax to make up for the damage they do to the environment, nuclear would be even more competitive. As fossil fuels get more expensive, as they must, the industry says that nuclear becomes even more attractive.

The industry also points to the recent disputes among Russia, the Ukraine, and other former Soviet states about the price of gas, which has caused temporary interruption to supply pipelines to Europe. Nuclear power would be a better bet, the industry says, ignoring the fact that uranium, while plentiful at the moment, is becoming more expensive. Prices rose 25% in the three months through December 2006, and a report by Resource Capital Research, an equity research company, forecasts that the price will rise another 75% in the next two years. Although at current consumption rates there is said to be enough reserves in Australia, Canada, and Kazakhstan to last 75 years, supplies are becoming more difficult to find.

While attacking fossil fuels, the nuclear industry correctly sees renewables as its main competitor for subsidies and capital. Scorning the newer technologies as being only minor players, they say that renewables will never be able to produce enough electricity to make up for the giant capacity needed—this can be provided only by building new nuclear plants. Wind turbines are particularly lampooned as being useless because the wind blows only some of the time. The industry then quotes the price of electricity produced by nuclear plants in pennies per unit, making the price look competitive with all other sources of power. The problem is that this price often reflects only part of the cost. In most cases it includes only the cost of mining the uranium, enriching it to make fuel, and the cost of running the plants while they are producing electricity. What is often omitted is the initial capital cost of building the plant, which is then written off, since this cost has frequently been provided directly or indirectly by state aid.

Another large cost factor is the disposal of the spent fuel and radioactive waste after the uranium

(continued on page 280)

Above: A view of the dormant cooling towers of the Unit Two reactor of the Three Mile Island nuclear power plant, as seen from the visitor center, near Middletown, Pennsylvania, in 1999, the 20th anniversary of the worst U.S. commercial nuclear accident. Although radioactivity was not released into the environment, there are many people who are still seeking an official admission that what happened at Three Mile Island was a meltdown. Whatever happened, it effectively stopped the U.S. nuclear industry in its tracks. President George W. Bush has tried to revive interest in nuclear power, but there is still little enthusiasm outside the nuclear industry.

"Nuclear power is too expensive and too dangerous. Wasting billions on more nuclear reactors would distract from the real task of developing renewables and reducing energy demand. Nuclear power is the ultimate unsustainable form of energy—for a little energy now, we would be condemning 10,000 generations to deal with the radioactive waste we would produce."

—Andrew Lee, director of campaigns
for the World Wildlife Foundation, UK, March 2006

Above: A general view of the Chernobyl Nuclear Power Plant in April 2006 on the 20th anniversary of the nuclear disaster there. Chernobyl was the world's worst nuclear accident, contaminating thousands of square miles in Europe. Although few people were killed immediately, many are still dying as a result of the radiation released.

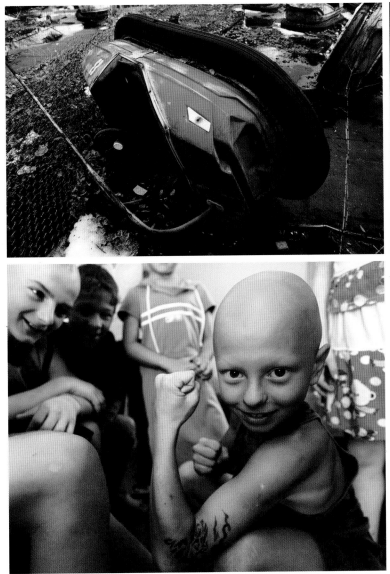

Top: A view of an amusement park in the center of the abandoned town of Pripyat, in the 19-mile (30-km) exclusion zone around the closed Chernobyl nuclear power plant, March 31, 2006. About 50,000 Pripyat residents were evacuated after the disaster, taking with them only a few belongings.

Bottom left and above: Official denials of the large number of children suffering severe illnesses in the wake of the nuclear accident at Chernobyl in 1986 have made people more cynical about nuclear power, not less. Since 1990 Cuba has treated 18,000 Ukrainian children free of charge for loss of hair, skin disorders, cancer, leukemia, and other illnesses attributed to the radioactivity unleashed by the meltdown years before they were born. These children do not appear in official statistics. Here, Bagdan, a Ukrainian boy, is surrounded by friends in a hospital at Tarara, outside Havana, in 2005. Natalia, also Ukrainian, receives treatment for alopecia (hair loss) in the same hospital outside Havana.

has been burned in the reactor. Ironically, in the past this waste has actually been counted as an asset since it could be reprocessed to recover spent uranium and plutonium, potentially another fuel. In fact, there are thousands of tons of this recovered material that have never been used and is now being stockpiled. This is a dangerous liability in this age of terrorism. Dealing with this material, which the industry calls "back-end costs," is either completely ignored or heavily discounted when calculating the cost of building new nuclear plants. That is because these costs will not have to be paid for many years and the burden will mostly fall on the next generation.

In reality, the cost of disposing of nuclear waste nearly always falls on the taxpayer. In the UK the part of the nuclear industry that was privatized would have gone bankrupt in 2004 if the government had not agreed to pay all nuclear-waste costs until 2086, at a cost of $370 million a year. That fact has been hidden from the public. One other point that has been largely ignored is that building new nuclear plants takes many years, usually more than a decade, from planning to producing the first electricity. In that time billions of dollars in capital are tied up as thousands of tons of concrete are poured and steel manufactured. This produces large quantities of carbon dioxide. Conservative estimates suggest that worldwide 1,000 nuclear plants would be needed to make a dent in the carbon dioxide problem. This raises the question of whether the industry has the capacity to build that many plants, or whether there is enough available uranium to go around.

In the first round of building nuclear plants, when sophisticated terrorism tactics and suicide bombers were the stuff of nightmares and not part of the horror of modern life, there was a thriving export industry in nuclear plants.

Russia, the United States, France, and the UK all competed to export nuclear plants to the parts of the world where providing nuclear know-how today would be regarded as a security risk. All four of these countries are still pushing the nuclear option hard and trying to sell the technology to any state they consider stable. In this volatile world, where a commitment to the safe operation of a new nuclear plant must be for at least 40 years, some of these judgments are highly suspect and have much more to do with power politics than a secure energy supply or climate change. Compare exporting nuclear technology, with its undesired dual ability of making atom bombs, with providing renewable technology. There are no security risks or waste problems in solar, wind, hydropower, and biomass.

But let us examine the funding alternatives for nuclear power. Take the capital cost for one nuclear plant, say conservatively $3.7 billion over 10 years, before one watt of electricity is produced. Experts have estimated that if the same sum were invested in energy-efficiency measures, particularly in the United States where so much energy is wasted, there would be no need for nuclear plants to be built at all.

But let us compare building nuclear power plants with renewables. For $4 billion hundreds of small-scale renewable alternatives could be up and running, some of them within months. Solar panels, small-scale hydropower, and wind turbines on homes, offices, and factories would give an instant return in terms of electricity and capital employed to the places where it was invested. The renewable options also avoid the problem of having nuclear plants located far from population centers because of safety reasons, which results in about 10% of the nuclear power generated is to be lost in long transmission lines before it reaches the customer. This is another cost the nuclear

Above: There are two models of the Japanese city of Hiroshima in the city's Peace Memorial Museum, which commemorates those who died when the U.S. Air Force dropped the atomic bomb in August 1945. The model shown at the top is of the city before the bombing. The model shown at the bottom is of how Hiroshima looked afterward. It was the first place in the world to experience the devastation caused by a nuclear explosion. The museum shows toys, books, and magazines that belonged to children at the time of the bombing. It is the fear of this kind of devastation happening again that makes some oppose nuclear technology.

Opposite, top: A warehouse containing the remains of spent fuel rods from a nuclear reactor. This waste can remain lethal to humans for 250,000 years. Every country that has adopted nuclear power is faced with the problem of what to do with the waste. This is the temporary nuclear-waste storage facility in the north German village of Gorleben, constructed to hold 420 containers of waste. The German government is hoping to find a permanent resting place for the waste in old salt mines, which are said to be very stable geological formations.

Opposite, bottom: A worker walks past the yet-to-be-filled nuclear-waste containers that will be stored in the warehouse. When the containers are full, it will be too dangerous for anyone without protective clothing to be this close to the waste.

industry fails to take into account or at least to admit to the public. The industry, faced with those arguments, says that countries can build both renewables and nuclear, which ignores the point that even in the richest countries there is not enough capital available for everything. To cut out the transmission problem, the Chinese are working on a small-scale reactor the size of a dustpan. This would provide local supplies for a village or small town. They also claim it is safe because it has an automatic shutdown mechanism if it overheats.

Angela Merkel, as her first policy act as the new chancellor of Germany, decided to spend her money on energy efficiency. Over a 20-year period all older German housing will be brought up to modern standards at the rate of 5% a year, until every home in the country is both warm and energy efficient. In one stroke she has created employment for thousands of people in the former East Germany, where jobs were in short supply, and eliminated the need to build a new nuclear plant. The old plants will become redundant.

The United States and Britain could adopt similar policies, but politicians in both countries seem to prefer building new power-generation plants. They show a preference for the giant and wasteful technologies of the last century.

An example of doomed but currently celebrated grand design is in Dubai, the small desert kingdom on the Persian Gulf. An astonishing 250,000 men, mostly from India and Pakistan, are building a forest of luxury hotels, roads, airports, artificial islands, and golf courses

believed to have a price tag of $100 billion. Dubai will be the business center and playground of the Middle East, and it is all being funded and fueled by oil. Golfers in Dubai used to play on sand, and there is still the celebrated old course, where golfers can still play in traditional style, without a blade of grass in sight. But it is surrounded by new lush green courses, with attendant country clubs, as if the desert no longer existed. In one sense it does not.

Desalination plants, using cheap oil from under the desert, provide the water for this lavish lifestyle. These plants use waste heat and surplus power from the electricity plants to produce fresh water. Here, in the driest part of the world, with sometimes less than 2 millimeters of rain a year, is the highest concentration of desalination plants in the world. More than 70% of the water in the United Arab Emirates, of which Dubai is part, comes from desalination— more than 6 million gallons (24 million l) a day. The country has the highest consumption of water per person in the world, with the exception of the United States. Much of the water is recycled, and sophisticated sewage works mean that water is not wasted, but this new "civilization" in the desert will almost certainly be one of the shortest lived in human history. It might be possible to build a nuclear power plant or, more sensibly, to use the plentiful solar energy now to replace some of the oil, but for how long? Well before the massive new towers and luxury homes reach the end of their lives, sea-level rise will overwhelm the low-level artificial island paradises, on which much of Dubai's new wealth depends.

Right: Germans believe that they can phase out nuclear power by adapting energy-efficiency measures and building giant wind turbines like this one at Brunsbüttel, near Hamburg. At the time this picture was taken, in October 2004, this was the largest wind turbine in the world, at 5 megawatts. In 20 years wind turbines have grown in size 10-fold, the earliest models producing only half a megawatt of electricity. This crane is lifting the blade hub 131 yd. (120 m) to the top of the mast. The span of the blade is 137 yd. (126 m). This is a prototype that is being built onshore, but RE Power Systems hopes to install large numbers in offshore wind farms in German waters and abroad.

What Can We Do?

Previous spread: These solar concentration dishes have been in operation from the early 1980s and are claimed by the makers Stirling Energy Systems of Phoenix, Arizona, to be the most efficient way of turning sunlight into electricity. These systems, which contain 89 mirrors, are made in sections in a factory to make sure they are in correct alignment and then transported on a truck to the site. They can be bolted together by two workmen in about four hours. Each morning the system "wakes up" and swivels itself into position to face the sun—turning like a mechanical sunflower to get maximum heat. The sunlight is concentrated by the mirrors to heat up hydrogen gas in the cylinders in the tower above. The expanding gas drives pistons at a steady rate of 1,800 r.p.m., about the same speed as a vehicle engine but in this case producing 25 kilowatts of electricity that can be used directly for local needs or put through a substation into the grid. With the price of oil remaining above $50 a barrel, the U.S. Department of Energy believes that thousands of these machines could make a major contribution to the electricity needs of the southwestern United States.

Above: These spectacular "blue marble" images are the most detailed true-color images to date that attempt to take in all of Earth from space. Using a collection of satellite-based observations, scientists stitched together months of observations of land surfaces, oceans, sea ice, and clouds into a seamless mosaic of every square mile of the planet. This image shows the Americas but with a tilt toward the North Pole so that the extent of the sea ice and the huge mass of the Greenland ice cap is clearly visible. It also gives a feeling of how vulnerable and small the planet really is.

Above: The colors on these globes above and opposite denote the type of landscape or the depth of the sea. It is startling how much of the land surface is desert or a type of prairie or grassland, and how little of it is green. It is also interesting to note how small Europe appears to be in relation to the other continents and the full extent of the Indian Ocean. Much of the information contained in these images came from a single remote sensing device, NASA's Moderate Resolution Imaging Spectroradiometer, or MODIS. Flying over 450 miles (700 km) above the Earth on board the Terra satellite, MODIS observes a variety of oceanic, atmospheric, and terrestrial features.

It is not surprising that individuals sometimes feel powerless when faced with really big issues like man-made climate change. What can one person do when the rest of the world seems intent on self-destruction?

Politicians do not help much with this. While they exhort individuals to take responsibility for their own actions, they do little to help by changing the regulations or tax system. Still worse, one of their constant excuses when asked to do something to show leadership on climate change is that one country cannot stand alone. George W. Bush took it one step further. His reason for repudiating the Kyoto Protocol was that the United States and the rest of the industrialized world would be disadvantaged against the developing world if they took action to curb carbon dioxide emissions, while others were allowed to pollute as they liked. Since the United States is the world's largest contributor to greenhouse-gas emissions, this was abdication of responsibility on a breathtaking scale.

For individuals the excuse used by politicians for not taking action is multiplied thousands of times. What difference can one person make when neighbors and most of the rest of the world carry on burning fossil fuels at an alarming rate? Of course, the honest answer to that question is that if everyone else did nothing, a personal effort would be futile. Stopping the global warming juggernaut would be impossible, even if having tried might make you feel better. Fortunately, there are two other incentives. The first is financial; taking personal action on climate change nearly always saves you and your family money. The second is that saving the planet is becoming fashionable, and burning unnecessary fossil fuels is increasingly being seen as antisocial. Soon it will be just as frowned upon as smoking close to babies.

Fortunately people acting to save the climate are not isolated anymore. All over the world people are learning to change their habits in hundreds of different ways to reduce their green-house-gas emissions. These individuals are business men and women, politicians, civil servants, shopkeepers, and all sorts of professionals and craftspeople. All of us are also consumers, making such decisions as whether we really need another plastic bag from the supermarket or a four-wheel-drive all-terrain vehicle to take our children one mile down a city street to school.

Every day it gets easier to make the right decisions because information about new products for households and better support for technologies that avoid fossil-fuel use are being made available. For example, 10 years ago long-life lightbulbs were large and clumsy appliances not suitable for many light fittings. Now they come in all shapes and sizes, are far cheaper, and are often indistinguishable from ordinary bulbs, except that they use only one-fifth of the power to do the same job and last 10 times as long. Forget for a moment the harm excess-energy use does to the planet and just consider the time and money long-life lightbulbs save you anyway. Anyone who does not use them is surely lacking in common sense.

For a whole range of home products, the level of information about energy use and education about the benefits of tackling climate change have also improved dramatically. At the city level, forward-looking mayors and city governments are transforming the quality of life for their citizens by providing better heating and lighting and reducing pollution and traffic, while at the same time helping the planet.

In writing this chapter, I am referring to a booklet called the "Little Green Book of Big Green Ideas," produced by Friends of the Earth, and to a compilation of 100 ways to save the

On this page: Road transportation is one of the largest and fastest-growing forms of pollution and perhaps the most difficult to control and reduce, but potential solutions, without losing the advantages of personal transportation, are developing fast. At top left bicycles in China, which has always had the largest number of cyclists in the world as well as the largest population. But in big affluent cities like Beijing and Shanghai, the bicycle is disappearing as more and more Chinese buy cars. Elsewhere in the world, especially in Europe, the bicycle is making a comeback, and ironically most of the bicycles are made in China. At top right California governor Arnold Schwarzenegger, who has done more than any other politician in the United States to stimulate cleaner transportation, inspects the Tesla Roadster electric sports car during his visit to the 2006 Los Angeles Auto Show. At bottom left is an electric motorbike by the Urban Mover company, one in a range of electric bicycles and scooters that are now the clean, cool, and cheap way to travel in cities. The motorbike has a maximum speed of 42 miles an hour and a range of 50 miles without recharging. At bottom right the Chevrolet Sequel, which General Motors claims is the most technologically advanced automobile ever built. It has a hydrogen fuel-cell propulsion system, aluminum body, fast acceleration, and a range of 300 miles (483 km).

Above: German policemen watch members of the environmental group Robin Wood unfolding a banner at Berlin's Brandenburg Gate at the end of March 2006. The group was protesting against the use of fossil fuels and atomic power just before a German government energy summit in an attempt to persuade the coalition of parties to phase out both as soon as possible. The banner reads: "Back to the rub?," "Coal kills climate," and "The future is renewable."

environment, produced by *Vanity Fair* magazine. A few years ago it would have been unthinkable for these organizations to be on the same wavelength. The two publications include advice about a range of disparate ideas to help the environment, not just about combating climate change but also about how to avoid using dangerous chemicals and eating endangered-fish species. Both are excellent, and if half the population had a shot at only half of the ideas, then the world would be dramatically changed for the better. Other organizations, like Greenpeace, have long believed that just telling people what the problems are, apart from being depressing, leaves them feeling powerless. That is why the organization had a solar demonstration plant mounted on the back of a truck that tours the United States powering concerts and other events. It also set up a wind-power company so that the public can invest in and buy green power.

But the first and easiest way of helping the climate and yourself is with energy efficiency. Some measures are so simple and obvious that it seems superfluous to mention them. But the fact is that most of us simply do not think about it and still less make a conscious effort to do anything about it. So below are a few of the most straightforward ideas.

The first and easiest way of saving energy and money is by turning off lights and machinery that are not needed. Televisions and computers are two examples. Every modern home is also equipped with an easily adjustable thermostat. Homes can be marginally warmer or colder at the flick of a switch, saving remarkable quantities of energy and bills at the same time. The same is true of offices. The first time I traveled to the United States at the invitation of the U.S.

administration, I was advised by the UK Foreign Office to take a sweater because it was summer and the air-conditioning in government buildings was so fierce I would be cold unless I wrapped up. The officials were right. The buildings in Washington were vast, and the cost of cooling them to well below a temperature necessary for comfort must have been immense—never mind the waste of energy.

One-third of all energy in many countries in northern Europe and North America is used in the home. This figure emphasizes the importance of cutting the consumption of energy in everyday life, but it is clearly a difficult area for politicians to prescribe solutions. They have largely left the public in limbo, merely urging them to be responsible but not helping much with what this means in practice. Unfortunately, the average American family doesn't stack up well on a world scale, producing more than 20 tons of carbon dioxide or related gases a year, compared to 4½ tons for the average resident of the planet. This is changing with the advent of home-energy audits. This involves inviting to your home a qualified person to assess your energy rating. Many utility providers or local governments will perform this service free of charge or for a nominal fee. Systems vary, but one program uses a score between 1 and 120. The higher the score, the better the quality of energy efficiency—with 1 being the equivalent of standing in an open field. By making some simple changes, the average family can eliminate 1,000 lb. (455 kg) of carbon dioxide emissions per year.

Since every house is different, with many built before the age of building regulations, remedies to improve your score, using any of these scales, come in all shapes and sizes. The auditor's job is to suggest the simplest and most cost-effective ways to improve the score, thereby reducing your

Top: German Chancellor Angela Merkel stands between Labor Minister Franz Müntefering (right) and Brandenburg's State Premier Matthias Platzeck as they pose for a picture during a cabinet meeting in the eastern German village of Genshagen on January 9, 2006. They and other politicians in Germany's coalition government, seen behind, had just clashed over the size of a major new jobs and growth initiative and the future of nuclear power on the eve of a special two-day cabinet meeting, and they wanted to show that despite this, the government was united.

Bottom: In 2006 factories in Germany produced roughly 66% more solar cells than during the previous year, spurred on by the reforms made by Angela Merkel. Here a man installs solar panels on a house, one of the many embracing the widespread domestic use of solar power.

energy bills. An audit should also tell you that if you spend, say, $200, how long it will take you to get your money back in reduced bills. Many of the remedies you can do yourself—for example, putting in attic insulation or fitting weather stripping in ill-fitting window frames. These are two of the simplest and most cost-effective means of saving energy and also the most widely neglected. It is astonishing how few people do these kinds of things, especially when there is so much public information available on the subject.

More complex operations, like adding wall-cavity insulation, clearly need specialized help. Almost without exception, where it is possible to install insulation in areas where there is none, it is very cheap, easy, and cost effective. The cost of both the insulation and installation can usually be recovered within a year through lower heating and cooling bills.

Windows are a different matter. They, too, have energy ratings, but the return on capital in terms of lower heating bills can take a lifetime. On the other hand, if you are replacing windows anyway because they need it, getting the best insulated type makes a noticeable difference in comfort without increasing costs greatly. There are also windows available with glass that reflects heat back into a room or helps to keep out harmful ultraviolet light.

Another area worth considering is passive heating and cooling, which can reduce energy bills by as much as 50%. The principles are obvious. Anyone who has a south-facing window knows how much warmer it can get on a sunny day. This can be a bonus in winter but make life uncomfortable in the summer. Making maximum use of this free heat gain from the sun in winter is what passive heating is all about. The simplest method is to expose or install a stone or tile

floor—the darker the better—where the sun shines through the window. This can be important in east- or west-facing windows as well. The floor will heat up during the day and release the warmth into the room during the evening. There are many variations on this method; these include building glass walls on the outside of a house and using the walls as passive radiators. The air between the glass and the wall also heats up. Small fans or vents at the tops and bottoms of the walls can circulate the warm air. Equally, a sunroom or conservatory can be used to trap heat using either a water feature or dark stones to store the heat for release later.

But, depending on your location, summer cooling can be as much a problem as winter heating. Here east- and west-facing windows and skylights can be a problem, particularly in the morning. The obvious solution is blinds or curtains, that remain drawn to keep the cool air in and the heat out. On south-facing windows, where the sun is high in the summer sky but welcome in the winter, adjustable awnings can solve the problem by allowing you to decide when the sun comes in.

Intelligent garden planting can make a difference, too. For example, deciduous trees or trellises with climbers can prevent the sun from shining in during the summer while allowing its warmth to penetrate in the winter. Careful thought in planting so that the area closest to the house is kept cool in summer can lower the indoor temperature considerably. This can be achieved by carefully arranging containers on a patio so that the heat of the sun is used for benefit in winter and deflected in summer.

Just as important are nights. Passive cooling at night involves opening windows or vents

Right: Iceland is a mass of volcanic activity, and the hot rocks are very close to the surface, giving its people the chance to produce power from this abundant energy source with no damaging emissions to the environment. This geothermal power plant is the first in the world to combine the production of hot water with the production of cheap electricity. It is owned jointly by the Icelandic treasury and the local people and provides power and heating to local homes and industry. It is also a tourist attraction, with a visitors center, and bathing in the mineral-filled plant's hot water, which is said to provide a cure for skin diseases.

upstairs and letting cool air in downstairs. But while passive heating and cooling would count toward an energy audit, the concentration is on saving costs in generating heat and cooling, whatever the condition of the house. When purchasing a heating or cooling system, look for the international ENERGY STAR symbol. It is a simple way for consumers to identify products that are among the most energy-efficient on the market. Choosing an ENERGY STAR product over another model could help reduce your carbon dioxide emissions and save hundreds of dollars in energy costs. In fact, in 2006 Americans, with the help of ENERGY STAR, saved enough energy to avoid greenhouse-gas emissions equivalent to those from 25 million cars—all while saving $14 billion on their utility bills.

Perhaps the simplest and most overlooked of all energy uses in the home is lighting. Lighting accounts for 30% to 50% of an average building's energy use. As I have already mentioned, long-life lightbulbs use much less energy to provide the same amount of light, but they produce far less heat. As a result, rooms stay much cooler in the summer. CFLs (Compact Fluorescent Lights), cost three to five times more than conventional bulbs but last up to 10 times longer and use one-quarter the electricity. If every household in the United States replaced one conventional bulb with a CFL, it would have the same effect on pollution levels as removing a million cars from the road. If you want to cut your bills further, install sensors that turn lights on when you come into the room and off again when you're not there.

Electrical appliances of all kinds—particularly televisions, computers, music and DVD players, and charging equipment for cell phones and other devices—should be unplugged when not in use. When a television is turned off but still plugged in, it continues to use energy, namely 25% of the energy it consumes when turned on. A simple solution would be to plug appliances into a power bar or strip, making it easy to switch them off.

Most homes are now full of electrical equipment, which is frequently either being added to or replaced. Throughout Europe, the United States, and Canada, on an increasing range of goods, there is a legal obligation for manufacturers to tell customers how much energy the appliance they are buying uses. In the United States and Canada, look for the EnergyGuide label, which estimates how much energy an appliance consumes, compares the energy use of similar products, and lists approximate annual operating costs. The most energy-efficient models will bear the Energy Star symbol, indicating that they incorporate advanced technologies that use 10% to 50% less water or energy. For the buyer it is best to go for the most energy-efficient model. What is also important to consider is the size of the refrigerator, freezer, washing machine, or other piece of equipment. A larger model, whether it has a good energy rating or not, is going to use a lot more power than a smaller machine. So consider what your requirements really are. A smaller model requires less space, is probably cheaper, saves money and energy, and might meet all of your requirements.

All of these small decisions about which sort of equipment to use can save you a lot of money and the planet a lot of carbon dioxide emissions, without altering your lifestyle in the slightest—except perhaps saving you the effort of changing lightbulbs.

There are lots of other simple methods to save energy and reduce bills considerably. Turn off radiators or space heaters in rooms that are not in use. If your home heating is not already

Opposite: A Chinese worker transports scrap metal to a recycling factory in Foshan in Guangdong Province in southern China in February 2006. He is one of an army of individuals and companies attempting to satisfy China's insatiable thirst for raw materials.

Above: Another worker attempts to capitalize on one of China's problems—in this case, waste. This load of plastic bottles and containers balanced on a single handcart are for recycling in Haikou, the capital of the southern province of Hainan. With one-fifth of the world's population, the amount of waste in China has become an enormous problem. About half of the waste plastic is left uncollected or dumped in an uncontrolled manner on land, in rivers, or in the sea, according to a 2003 collaborative study by the country's Institute for Environmental Studies and the Chinese Academy of International Trade and Economic Cooperation.

> "The impacts of climate change will fall disproportionately upon developing countries and the poor persons within all countries. It will therefore exacerbate inequalities in health status and access to adequate food, clean water and other resources."
>
> —Rajendra Pachauri, chair of the Intergovernmental Panel on Climate Change, July 2005

controlled by a programmable thermostat, have one installed. This low-cost device enables you to automatically adjust the temperature to your requirements resulting in lower heating bills and reduced carbon dioxide emissions. There are dozens of other small tips that are both obvious and easy to forget—for example, not filling the kettle too much when you boil water for tea or coffee. Kettles are now on the market that will boil exactly the amount of water you need for one or two or more cups, saving both time and energy.

Outside the home there is also much that can be done. Almost one-third of the carbon dioxide produced in Canada and the United States comes from cars, trucks, planes, and other vehicles that deliver food or products. Again there are options that can save money and will not change your lifestyle, and there are those that make a statement of intent. The best example, perhaps, is the kind of vehicle you choose to drive. Almost all new cars, irrespective of engine size, can easily travel up to 70 m.p.h. (10 kph), which is at or above the normal top speed allowed by law. However, a vehicle's fuel efficiency drops off above 55 m.p.h. (90 kph), so observing the speed limit actually reduces greenhouse gas emissions and saves you money. Keep your car in good running order. Replace clogged air filters, ensure that your tires are aligned and fully inflated, and keep the engine tuned. Doing these things could save the average two-car family more than $1,000 in gas costs over the course of a year, as well as reducing greenhouse-gas emissions.

Purchasing more fuel-efficient vehicles is another way to save money and reduce emissions. Governments in Canada and the United States frequently tax gas guzzlers at a higher rate, while offering financial incentives to buy or use energy-efficient vehicles. Check out whether you may be eligible for a tax credit for the purchase of a hybrid vehicle.

Following Hurricane Katrina, the lineup to buy the dual electric/gasoline Prius was in sharp contrast to the inability of Ford and General Motors to offload their extremely low-miles-per-gallon SUVs. And automakers increasingly see hybrids as the way of the future. To cite just a few examples, recently Toyota announced that it expected hybrids to account for 100% of its new cars by 2020, and Chrysler has stated that its Hemi V-8 engine—the one used in some of its most thirsty trucks—will be getting a hybrid makeover for 2008.

All sorts of new technologies, including cars driven by hydrogen-powered fuel cells and even compressed air, are under development. Electric scooters are now on sale worldwide. As can be seen elsewhere in this chapter, there are lots of cleaner and greener ways of getting around than with the old-fashioned gasoline-driven car. Ultimately, the idea is to create a system that allows a car owner to gain tomorrow's "fuel" by plugging in his vehicle to an electric socket when he goes to bed at night. Whatever technology the driver has chosen, the car will be recharged with electricity, restocked with hydrogen, or filled with compressed air in time for the next morning's drive to work. While the world waits for this new generation of vehicles to arrive, it is still easy to save money and reduce your current emissions by making the next family car as fuel efficient as possible. If bio-diesel or ethanol is available, even as part of a fuel mixture, use it. Most cars are capable of running on E10, a mixture of 10% ethanol (made from corn) and 90% gasoline. Already half of the almost 11 billion barrels of corn the United States produces each year is converted to ethanol. Even better, researchers are working on alternatives that don't rely on food crops, such as municipal waste, used cooking oil, wood pulp, and leftover grain. It is, of course, often less stressful to leave the car at home and travel by public transportation

Right: For the people who work at this plastic recycling depot in Bangladesh, recycling is a job that provides them with money to live. There are thousands of families for whom earnings from recycling by hand is a welcome alternative to no income at all. The millions of bottles have to be retrieved from the trash, sorted into colors, cut up, and washed in the Buriganga River before being recycled.

"Every action or inaction has an impact—
good or bad—upon our surroundings, and
anything we do today will have an impact
on the lives of our grandchildren."

—Ted Turner, Chairman's Letter, Turner Foundation Inc.

whenever possible. It is also healthier and cheaper to walk or ride a bicycle, particularly for short trips.

Lots of other personal choices in our daily lives can also make a difference. One example of this is what we eat. Over the last 20 years it has become possible to go to the local supermarkets and buy virtually any meat, fish, fruit, or vegetable in any season. Many of these items are flown across the world to meet the whims of shoppers for out-of-season strawberries or peas. These so-called food miles are adding greatly to environmental costs, especially air flights—which have increased 23% in 30 years—and supermarkets are largely to blame.

This could have disastrous consequences for all of us. As well as increasing greenhouse gases, local farmers, who cannot sell their products because of cheap food imports, are put out of business. This is usually not straight competition. Cheap labor or near slave labor in the developing world together with massive farm subsidies in Europe and in North America distort world prices. The result of all this has been the loss of a tradition: local produce for local needs.

At last, however, the consumer has revolted, or at least enough of them have that attractive alternatives to supermarket shopping are springing up. Farmers' markets are gaining in popularity throughout Europe and North America, offering locally grown produce. Another development is community-sponsored agriculture (CSA), where farms, often organic farms, deliver locally grown produce to your door each week. Again it is direct selling of local produce in its season. This might seem a departure from the topic of climate change, but as the price of fossil fuels rises and the land

for growing food is lost to sea-level rise, drought, and deserts, how food and other goods are produced is important. Fair trade and sustainable production methods—cotton is an example—are now being adopted by major stores as part of a successful marketing strategy to increasingly choosy consumers.

One of the great problems in dealing with man-made greenhouse-gas emissions is the issue of flying. All over the world, air travel is growing, and with it emissions high in the atmosphere, where scientists believe the pollutants magnify the problem. Some believe that the quantity of aircraft contrails, or vapor trails in the sky, actually masks the effects of global warming, as we saw when North American air travel stopped after the destruction of the World Trade Center. This might be an argument for allowing the increase in air flights to continue while we try to solve some of the environmental problems on the ground, but it is not one I would support.

Telecommuting for business when possible and taking vacations closer to home would help. Consider that trains are at least twice as energy efficient as planes. If a flight is unavoidable, you might reduce your footprint by buying carbon offsets to compensate for the emissions caused by your air travel. Optimists believe that the problem of air-travel emissions will begin to take care of itself. Since there is no alternative to fossil fuels for powering aircraft, the cost of escalating oil prices will immediately feed through to ticket prices. Air travel and vacations in distant places will begin to slip back into being an occasional luxury rather than the cheapest option for weekend breaks. People will vacation within driving or train distance of their homes.

Opposite, top: Solar panels provide the roof and generate approximately 250,000 kilowatt hours of electricity on Coney Island's Stillwell Avenue in New York City. This is the largest aboveground station in New York's subway system, and the semi-transparent photovoltaic panels were ideal for the 76,000-sq.-ft. (7,060 sq.-m) steel-and-glass structure, which adds beauty to the convenience of the railway system.

Opposite, bottom: Architects everywhere are trying to make their buildings practical and environmentally friendly while still being beautiful. The goal of sustainable development is often hard to achieve, especially when building airports. Shown here is part of what Malaysia says is the world's largest airport, with five runways, 35 miles (60 km) south of Kuala Lumpur. It connects to the capital by train, and there are internal driverless trains to take passengers between terminals. To offset the environmental costs, a tropical rain forest was transplanted from elsewhere to surround the airport, in addition to vegetation that was planted in the central garden space of the satellite area of the development.

Nearly all of the suggestions so far, apart from limiting your flying, come into the category of helping to save money and the planet without altering your lifestyle in the slightest, except perhaps to make yourself more comfortable. Other more ambitious efforts do involve making a bit of a statement as well as an additional outlay of cash. This includes fitting renewable energy to your home in the form of solar heating, photo-voltaics, or wind turbines. So far in most countries one of the problems of tackling climate change is that these new technologies need massive uptake to achieve economies of scale. Computers and cell phones are two examples of how this happened in other fields when big business got behind the marketing and selling of them. The price of renewable energy will fall dramatically when home-energy kits are produced like refrigerators or washing machines, and that is beginning to happen.

That is because in North America and throughout much of the world, there are other factors apart from people power that are pushing along the changes, particularly in dealing with climate change. There is increasing comprehension among governments that fossil-fuel prices are going up and will continue to do so because supplies are not meeting demand. The only solution, for all countries, if their economies are not going to be damaged, is to reduce fossil-fuel use, and the only way to do that is to develop alternatives.

For individuals, apart from a strong desire to reduce the dangers of global warming both for future generations and for ourselves, there are other incentives. For the first time the cost of energy is beginning to make people think twice about the sort of car they buy and how much they need to heat and cool their homes. North Americans use two to three times more energy per capita than other highly developed countries like France, Germany, England, and Japan. Although Canadians sometimes console themselves by pointing out that they produce only 2% of the world's greenhouse-gas emissions, they are one of the worst offenders per capita for energy consumption.

One way of making a contribution is to buy green electricity. At least 50% of electricity customers have the option to purchase renewable electricity directly from their power supplier. All customers have the option of purchasing renewable energy certificates that allow you to contribute to the generation of clean, renewable power, even if you can't buy it directly.

Generating your own electricity or hot water at home is a feasible option almost anywhere in the world and is becoming more economically viable every year. To encourage the trend, the U.S. Energy Policy Act recently implemented a 30% tax credit for installing solar water heating systems. Over much of Europe and North America, micro-wind turbines plugged into the home energy supply are also becoming a sensible option to draw free electricity from the elements. There are already a number of products on the market, and these are getting better all the time. Governments are slowly altering the regulations to make this a more viable option.

Those lucky enough to have access to a river or a stream could try mini-hydropower. Since technologies like solar power have a 20-year guarantee with them and will last far longer, many people regard the investment in renewables as a form of self-reliance, a sort of pension fund of free power. There are other ways of gaining energy from the environment like ground-source heat pumps. This is a way of extracting the Earth's natural heat by using a

Above: Swiss Re's London headquarters at 30 St. Mary Axe, known to almost everyone as the gherkin, opened in May 2004 with the boast that it would use half the power of a normal skyscraper. It has a series of shafts that cool the building in summer and distribute passive solar heat in winter. It was designed by the architects Foster and Partners to be London's first environmentally sustainable tall building. Among the building's most distinctive features are its windows, which open to allow natural ventilation to supplement the mechanical systems for a good part of the year. Wherever possible, recycled and recyclable materials were specified throughout the building. Swiss Re, the reinsurer, has done a lot to draw attention to the perils of climate change and commissioned the building to demonstrate that even distinguished buildings could also be environmentally sustainable.

Opposite: New York's 48-story building at 4 Times Square, designed by Fox & Fowle Architects, was one of the first buildings of its type to adopt standards of energy efficiency, use sustainable materials, and employ responsible construction techniques. Although it cost more to build than a conventional building, the owners had no trouble finding occupants because of its green image, lower operating costs, and better internal environment, as well as its location. The windows allow daylight into the building but keep ultraviolet light out while decreasing heat loss in the winter. Thin-film photovoltaic cells in the top 19 floors of the building supplement electricity supplies. Two natural-gas fuel cells provide 100% of nighttime electricity and some hot water and heating for the building. Special air intakes and filtration systems reduce the intake of pollution, the air quality is monitored floor by floor, and cleaning materials are nontoxic.

heat pump to raise the temperature to warm your home. Currently the payback period for a ground-source heat pump does not justify the capital cost unless you intend to stay somewhere for a long time or have one installed with new construction. Having your own source of electricity or heating is, however, a good selling point for a property.

But it remains true that rising energy prices are the greatest driver for change. It has been said many times, but it is worth repeating, that doing something about climate change is a massive boost to the economy rather than the opposite, as President Bush and the fossil-fuel lobby claim. This boost applies to individual households as well as nations. It is bizarre that governments, charged with looking after the public welfare, acknowledge this but still seem incapable of putting policies in place to take advantage of it. Finally, however, we are seeing the seeds of change.

Germany is a shining example of what can be done. In 1998 Germany began a 100,000 Roof Program, which gave people 10-year loans at reduced interest rates to buy photovoltaic systems. In five years the target was reached, and Germany had a competitive solar power industry. As well as boosting solar power, the German government at the same time embarked on encouraging wind power and leads the world in installed capacity. More recently, as reported in the last chapter, the government decided to bring all of its older properties up to modern energy efficiency standards at the rate of 5% a year. In 20 years all of the older housing, mostly in the former East Germany, will be brought up to modern standards. This is a classic example of how jobs can be created where they are most needed, housing stock improved, and fossil-fuel imports reduced, all at the same time. Reducing Germany's greenhouse-gas emissions and avoiding the need to replace

Left: The UK Fibropower station in Suffolk was designed by the architects Lifschutz Davidson Sandilands in the early 1990s to look as much like a grain silo as possible to fit into Suffolk's farming landscape. The Fibropower generating plant was the world's first to be fueled by poultry litter and now produces electricity for 12,500 local homes. Nothing goes to waste. At the end of the process, the raw material is turned into nitrate-free fertilizer, ready to be returned to the land.

the country's aging nuclear reactors are two other advantages. Germany has the good fortune to have one of the most environmentally aware populations in the world. The result is that individual voters are helping to drive the changes by buying into the new technologies in large numbers.

That is also true in California. The state's $2.8 billion Million Solar Roofs Program takes a similar approach to Germany's. It mandates that new home buyers must be offered solar panels as a standard option and offers cash incentives on solar systems. These incentives, combined with federal tax incentives, can cover up to 50% of the total cost of a solar-panel system. The goal is to create 3,000 megawatts of new solar-produced electricity by 2017, moving the state toward a cleaner energy future and making solar power mainstream.

While some technologies require government incentives to create a market volume that makes them economic, others are already economic but underused. An example is solar water heating, which in some countries, like Turkey, has long been a standard way of providing hot water for family homes and hotels. These solar systems do not produce electricity but simply heat water on a rooftop and circulate it into a tank. A 2-square meter panel on a family home can reduce annual water heating bills by 70% in sunny countries, with the southern United States and the Mediterranean area being examples. Solar water heaters make economic sense without requiring any government intervention.

However, some populations still need a push to change their habits. Spain, which has the vision to create jobs by becoming a leading manufacturer of solar-tube systems for heating water, made them compulsory in 2005 for all new buildings, both domestic and commercial, substantially

reducing the nation's energy bills. China, which already has the largest installed capacity of solar water heaters in the world, plans to quadruple it in 10 years.

But while governments have an obvious role to play in regulation and tax incentives to stimulate a mass market and to provide business confidence to create a manufacturing base in all of these technologies, there needs to be cooperation and creativity across society.

Developers and the construction industry have been slow to adapt to new building techniques unless driven by regulation, frustrating a new breed of architects who design houses that need little or no energy to make them comfortable. While it is possible to retrofit solar and water heating panels and mini-wind turbines, nearly all of the technologies are cheapest to install in new buildings. Increasingly popular on office blocks and private homes are green roofs. These provide roof gardens and insulation, reduce flash flooding, and encourage birds and butterflies. In some more ambitious schemes water from showers and washing machines can be recycled via a roof garden, both growing plants and purifying the water at the same time. A reed bed on a roof can be both a wildlife haven and a water-cleaning system at the same time.

These and other technologies need to be built into both commercial and domestic homes in the planning stage to gain the most benefit. Many local municipalities are already realizing that developers will not do this on their own because many of the innovations, while cutting the cost of running buildings, add to the initial construction costs. As a result, at the 2006 Atlanta Energy and Environment Summit, 120 mayors, business leaders, developers, retailers,

Top: Honda is another Japanese vehicle manufacturer that realizes the potential of hybrid technology. Already the hybrid system is in use on motorcycles and trains as well as cars. All of the world's major manufacturers are now working on models of their own.

Bottom: Cars that run on hydrogen are often billed as a solution to the twin problems of pollution and climate change. When hydrogen is burned in fuel cells, the only waste is pure water. This technology is no distant dream, because over 50 million tons of hydrogen are produced and consumed every year and fuel cells

work efficiently. But as with other new ideas, the problem is switching to a different world, where motorists can feel there is a practical alternative to buying gasoline and diesel vehicles and have confidence that when they drive into a gas station, there will be the alternative fuels they need. Shell has a global hydrogen business and is committed to moving production and sales from an industrial setting into the lives of ordinary people. Widespread use of hydrogen could also help address concerns about energy security. This is Shell's hydrogen-fuel retail site at Benning Road in Washington, D.C.

"It [the Toyota Prius] has become an automotive landmark: a car for the future, designed for a world of scarce oil and surplus greenhouse gases."

—Alex Taylor, *Fortune*, February 2006

Above: The 2004 Toyota Prius made its world premiere at the New York Auto Show on April 16, 2003. The Prius is a "full-hybrid system" car that can operate in either gasoline or electric mode as well as one in which both the gasoline engine and the electric motor are in operation at the same time. The battery that provides the electric drive is charged every time the motorist brakes. After

Hurricane Katrina struck in 2005, fuel prices rose and climate-change awareness grew in the United States, causing dealers to start waiting lists for the Prius because demand was so great. The car boasts as much as 50 miles per gallon in city traffic.

Above: This partly underground house was built in central London by pioneer architect Alex Michaelis for his own family. Planning restrictions meant the property could not be more than 6 ft. (2m) above ground to avoid blocking out the neighbor's light, so it has a flat roof, grassed over for added insulation. The water supply comes from a deep borehole below the house that also feeds into a heat exchanger, which provides heat in winter, cooling in summer, and warms the water in the family swimming pool in the center of the house. Electricity comes from Good Energy, a company that sells only green electricity with supplementary supply from solar panels. Alex's wife, a pediatrician, charges her electric car overnight for hospital visits the next day. Electric cars have free parking in London and are exempt from congestion charges.

Above right: A five-bedroom house in Oxford, England, that generates all of its electricity requirements from the 48 solar panels on the roof. Greenpeace believes that solar energy combined with proper construction is a realistic choice for all of Britain's electricity consumers because these panels produce electricity even on cloudy days.

and architects from across the United States joined forces to unanimously adopt a resolution calling for all new buildings in the United States to be carbon-neutral by the year 2030.

To force builders to employ forward-looking architects who have cut their environmental teeth on individual designs for discerning clients, planning authorities in Europe are imposing conditions on the mass market in order to cut emissions. An increasing trend in many European countries is to refuse planning permission for new housing schemes that do not have at least 20% renewable energy technologies built into their designs.

Boston now requires green construction for all private buildings that cover more than 50,000 sq. ft. (4,650 sq m). Of course, it is also in the interests of private home buyers to inquire what, if any, steps have been taken to protect them from the excess use of fossil fuels, which are set to continue rising in price. All of the evidence from across the world is that environmental benefits built into home design are a selling point. European law now requires an energy-efficiency assessment of every house sold and details of energy bills.

In most countries there are demonstration projects, and frequently there are whole suburbs in developed countries, designed to save energy and reduce environmental impacts. Many have become tourist attractions. China, which has a mass urbanization program, has gone one step further. It is preparing to build a whole city as an eco-project. With millions of Chinese now aspiring to city life in a Western style, they are planning to use renewable energy and combined heat and power plants for an eco-city in Dongtan, near Shanghai. The Chinese government is hoping that the city will be without the water shortages,

air pollution, and waste problems of many of its other rapidly developing urban centers.

Where it seems impossible to change your lifestyle or, in the short term, the way you run your business, it is possible to buy carbon offsets. This is another growing business across the world and potentially a great benefit both for green technologies and for forests.

The idea is simple. You calculate how much carbon dioxide is released into the atmosphere as a result of your activities; then you pay a company to invest in projects that save the same amount of carbon elsewhere. The Calgary Flames hockey team recently purchased carbon offsets to compensate for estimated greenhouse gas emissions associated with flights and hotel accommodation for their "away" games. Similarly, when you book flights with travel companies like Expedia or Travelocity, you can pay extra to offset your share of the flight's emissions. The money goes to a range of initiatives at home or abroad. A sample project might be funding a small-scale hydroelectric power station in India, which would provide carbon-free power and light and stop the use of inefficient, expensive, and polluting kerosene heaters.

The first and most popular form of carbon offsets in both Europe and North America is to plant trees. A project could be in the home country, producing new native woodlands, or replacing felled tropical woodlands in some far-off continent. In every case the point is that while the tree grows, it takes carbon dioxide out of the atmosphere and fixes it in the wood. In some cases just paying to preserve a forest that would otherwise be felled is regarded as an option. Other land-use projects that prevent carbon from being released from the soil can also qualify. The problem with these plans is that

"It is our task in our time and in our generation to hand down undiminished to those who come after us, as was handed down to us by those who went before, the natural wealth and beauty which is ours."

—John F. Kennedy, March 1961

the science of exactly how much carbon is saved is not clear-cut, and unscrupulous operators have made spurious claims and are even selling off newly planted woodland. Woods can also be cut down, catch fire, or be "harvested," releasing the carbon back into the atmosphere. This is not said to discourage people from planting trees. Forests certainly reduce the temperature, and woods also help retain water in dry areas, actually producing rain clouds of their own, while moderating the climate.

As was previously mentioned, renewable energy projects are a good option for carbon offsets, with solar, wind, and biomass as well as small-scale hydroelectric all qualifying. All of these have the benefit of allowing communities to develop without going down the carbon-intensive route.

NativeEnergy, a U.S. carbon offset company, recently helped fund an anaerobic digester on a Pennsylvania farm that harvests methane gas from cow manure to produce 7.2 gigajoules (2,000 kilowatts) a day of power. That is more than three times the amount of energy that the farm currently uses, enabling it to sell the surplus back to its power provider. Another project by Climate Trust will enable schools in Montana to convert steam boilers to burn wood pellets (made of waste products from local mills) rather than fossil fuels.

Another variation of carbon offsetting that is gaining popularity with businesses as a way to "do the right thing" by the environment is emissions trading. The Chicago Climate Exchange opened in 2002, providing the first market for emissions trading in North America. Companies that sign on agree to reduce their emissions by 1% per year. If they are unable to do so, they can purchase allowances by funding large-scale projects that reduce greenhouse gases.

We must be aware that not all carbon offsets and renewable-energy certificates are created equal. "There's a lack of standards in the voluntary market, and the offsets that people are purchasing might not be accomplishing what they hope they will," according to the David Suzuki Foundation, a Canadian environmental advocacy group. Look for offsets that have been independently verified by a third party and that clearly state which projects they support and how they are making a difference.

Even without national leadership, some 200 U.S. cities are planning to meet Kyoto Protocol targets for reducing emissions. And at the state level, California has called for a 25% drop from 1990 emissions levels by 2020 and 80% by 2050. Even Texas has ruled that 2.2% of its energy must be drawn from renewable resources by 2009. Businesses are joining the effort as well. IKEA U.S., for example, has begun charging customers 5 cents per plastic bag. It expects to cut bag use by 50% to 35 million, and the earnings from selling plastic bags will go to American Forests, a nonprofit conservation organization.

If it is not possible to buy "green" electricity, that is, stipulating that all the electricity you buy comes from renewable resources—such as wind, solar, hydroelectric, or tidal, for example— then it might be practical to generate your own, either individually or as a group. However, producing your own energy requires some thought. It might not be sunny or windy enough in some places for renewable energy to be an economic option, but there is bound to be at least one technology, which has already been developed, that is available where you live.

Opposite, top: The depletion of fish stocks around the world has led some chefs and supermarkets to insist on buying only from sustainable resources. Increasingly, fishing organizations are realizing that the past hunt for short-term profits has been killing their industry. This view of fishermen at sea with seagulls overhead is fast becoming a memory in some areas where fishing is banned because of over-fishing or where stocks are so depleted it is no longer worth putting out to sea. In the UK the Sea Fish Industry Authority now works across all sectors of the seafood industry to promote quality, sustainable seafood to try and keep the industry alive.

Opposite, bottom: In the UK and across Europe, there has been an explosion of interest in organic farming and markets where farmers sell local produce. FARMA, The National Farmers' Retail and Market, is a cooperative of people selling on a local scale. FARMA works throughout the UK and is the largest organization of its type in the world, representing direct sales to customers through farm shops, pick-your-own schemes, farmers' markets, and home delivery.

Innovative builders across North America are turning to green building techniques to reduce ongoing costs and minimize their impact on the environment. In New York City the Bank of America Tower will save "gray water" from washroom sinks to flush toilets, generate much of its own electricity from natural gas, and use a combination of concrete and slag (waste from blast furnaces) for building in order to leave a lighter carbon footprint. The Gulf Islands National Park Reserve in Sidney, British Columbia, uses ocean water to provide heat, the sun to create the bulk of its energy, and low-flush toilets that take advantage of plentiful rainwater.

In this chapter are just a few ideas of what you as an individual can do. There are many other good environmental ideas that will help as well. For example, recycling, composting, and not using the plastic bags offered by supermarkets all have a climate-change element as well as saving resources. One area that we have not yet touched upon is voter power. Many of the technologies mentioned above are held back because politicians have failed to change the tax system or create changes in the regulations to favor a cleaner way of life. Countries that have had the courage to ban or tax plastic bags have made a substantial environmental gain, but most governments have given in to pressure from supermarkets to continue the practice of giving bags away for customer convenience. The long-term detriment to the environment continues to be ignored. Voters need to fight back.

There are dozens of other examples where enlightened governments, urged on by consumers and voters, can change habits and markets. In doing so, they can literally change the world, the technologies that power it, and probably save the planet (or at least most of it) from impending disaster. It is a daunting challenge, but we can all help to do it.

Right, top: Children are the ones who will suffer most if we continue to alter the atmosphere and damage the natural environment. Even the youngest of students are concerned for the future. Here on June 6, 2006, in Panama City, students are marking World Environment Day. The banner reads, "Conserve our environment."

Right, bottom: Turtles are one of the oldest creatures on the planet, but in many places they are endangered because of interference with the beaches where they bury their eggs to hatch. To help them survive in the 21st century, turtles are protected in many countries until they are large enough to survive in the wild. Here Colombian Wayuu Indian students release turtles at Colombia's Camarones Beach on World Environment Day in June 2006.

Hydrogen. Iceland has ambitions to become the world's first hydrogen economy by 2050. It is a country full of natural resources to facilitate this change. Only 5% of Iceland's potential for geothermal energy is used at present, and there are also large resources of hydroelectric power.

Hydroelectric. Water power is perhaps the oldest form of harnessing renewable energy. Today kinetic energy from water movement creates modern hydroelectric power. There are valid arguments questioning its efficiency on a large scale, but small-scale schemes have also been embraced in countries such as the UK, Georgia, Armenia, China, and India.

Carbon trading is a complex concept, revolving around the buying and selling of credits for reducing pollution. It means that industries can gain cash by investing in energy-efficiency measures. It is a changeable market, but at a high in April 2006, the tons of carbon saved by industry was potentially worth $80 million.

Energy at home. No matter how large the difficulties facing us may seem, small adjustments in our daily lives will make a difference. Fitting insulation, using energy-efficient lightbulbs, and appropriate water management are measures we can, and should, all take. In the UK organizations such as the Carbon Trust are dedicated to educating and facilitating these changes.

U.S. individual mayors. Despite the often disheartening response to the threat of climate change by the federal government, the actions of individual states within the U.S. shows all is not lost. Among such good examples are the mayors of Seattle, Chicago, Miami, and Albuquerque who recently proposed Resolution No. 50, setting a goal for carbon neutral buildings by 2030.

Ordinary people, governments, and companies all around the world are using new technologies coupled with common sense to combat climate change. From gargantuan wind turbines to the small changes we can all make in our daily lives, it is obvious that we have the tools and know-how to solve this problem. All that is required now is our commitment to act.

Biofuels. In Brazil more than half of the cars sold can use ethanol for fuel, burning 4 billion gallons annually. Oil prices in the 1970s prompted the development of this by-product of sugar into a potentially huge earner for Brazil as demand for cleaner energies boost their exports.

Tidal turbines. The world's first commercial-energy wave farm is being constructed off the shores of Portugal. This technology generates renewable electricity from ocean waves, using both tidal turbines and sails to capture the force of the tide and may prove to be a valuable energy source for the 21st century.

Wind power. Denmark leads the world in wind power—over 20% of electricity consumption in this Western developed nation is covered by energy from wind turbines. This is the equivalent of 1.4 million Danish homes. Fiscally, the wind industry brings 3 billion euros a year to the Danish economy and employs over 20,000 people.

Reforestation. In 2002 China embarked on a 20 billion yuan (over $2.6 billion) 10-year plan to reforest barren lands. About 170,000 sq. miles (440,298 sq km) have been earmarked for reforestation. These efforts will extend to attempting to create barriers to shield cities such as Beijing from sandstorms.

Ocean thermal-energy conversion. Japan, in a bid to combat its dependence on others for oil, began research into alternative energy as early as 1950. Ocean thermal-energy conversion is an example of this forward thinking, a technology with the potential to supply fresh water, hydrogen, and lithium, as well as electricity.

Hybrid vehicles. Japan's automobile manufacturing industry has the headstart on hybrid vehicles, with Toyota and Honda leading the way. The Toyota Prius is the number one–selling hybrid car in the world, with owners ranging from Hollywood stars and senior world politicians to ordinary environmentally conscious people everywhere.

Carbon storage is the process of removing carbon dioxide from the atmosphere and pumping it underground so that the buildup of carbon dioxide concentration in the atmosphere will reduce or slow. This technology, also known as carbon sequestration, is available globally, most recently in the Sahara, already suffering the effects of climate change.

Solar technology, as demonstrated on this Greenpeace tour, has spread throughout the world, an example of the basic joy of renewables in that wherever the sun shines, energy can be found. Solar provides pollution-free electricity where it is installed without the need for a grid. In India the Tata Group and BP set up Tata BP Solar Ltd, with the aspiration to improve quality of life throughout India and the world with clean energy.

Geothermal technology utilizes the natural heat of the earth to provide energy. In the Philippines geothermal energy represents 27% of the country's total electricity production and is second only to the United States in global geothermal energy production. Development started as early as 1977 on the island of Leyte and continues today.

Index

Photo Credits

Front and back cover iStockphoto

Front matter
2 iStockphoto
5 National Geographic/Getty Images
6-7 iStockphoto
8 Menn/Greenpeace

Chapter 1
10-11 Craig Mayhew and Robert Simmon, NASA/GSFC
12 European Space Agency. All rights reserved.
13 Jacques Descloitres, MODIS Rapid Response Team, NASA/GSFC
14 Jacques Descloitres, MODIS Rapid Response Team, NASA/GSFC
16-17 Reuters/Ivan Milutinovic
17 Reuters/Enrique Marcarian
17 Reuters/Eduardo Munoz
19 Tongjing/Greenpeace
19 Reuters/China Photo
20 Greenpeace/Steve Morgan
21 Michael Van Woert, NOAA NESDIS, ORA
21 Michael Van Woert, NOAA NESDIS, ORA
22-23 GOES-12 1 km visible imagery, NOAA
24 Reuters/Brian Snyder
24 Reuters/Larry Downing
25 Reuters/Brian Snyder
26 Reuters/Carlos Barria
26 Reuters/HO Old HR/RC/ME
26 Reuters/US Army/Staff Sgt. Suzanne Day/Handout RCS/KS
27 Reuters/Eriko Sugita
28 Reuters/Mike Blake
28 Amjad Abdulla
28 George Fischer
28-29 NASA/GSFC/ASTER Science Team
29 Brian Knutsen
29 George Fischer
29 Brian Knutsen
31 Reuters/Rafiquar Rahman
31 Reuters/Seth Wenig
32 Greenpeace/Bond
32 Greenpeace
33 Langrock/Greenpeace
34 Shirley/Greenpeace
34 WWF-Canon/Mario Farinato
36-37 Image by Robert Simmon, based on IKONOS data provided by Space Imaging through the NASA Scientific Data Purchase, with help from Danielle Grant and Compton Tucker, NASA/GSFC

Chapter 2
38-39 Reuters/Simon Kwong
40 MODIS Rapid Response Team NASA/GSFC
41 Reuters/Rickey Rogers
43 Reuters/Jose Manuel Ribeiro
44 David Dodge/The Pembina Institute
46-47 Reuters/Andrew Wong
50-51 NASA/Goddard Space Flight Center Scientific Visualization Studio
52 National Geographic/Getty Images
53 Reuters/Ray Stubblebine
54 Jacques Descloitres
55 Greenpeace/Beltrá
57 TopFoto/ImageWorks
59 Image provided by the USGS EROS Data Center Satellite Systems Branch

Chapter 3
60-61 Reuters/Georges Bartoli
62-63 NASA/Goddard Space Flight Center Scientific Visualization Studio
65 Reuters/Shamil Zhumatov
65 Reuters/Shamil Zhumatov
67 Reuters/Phillipe Desmazes
68-69 Jacques Descloitres, MODIS Rapid Response Team, NASA/GSFC
70 Reuters/Christine Muschi
70 Reuters/Christine Muschi
70 Reuters/Michael Dalder
72-73 Greenpeace/Andrew Male
74-75 Reuters/Gregg Newton
77 Jacques Descloitres, MODIS Rapid Response Team, NASA/GSFC
77 Topham Picturepoint
77 Greenpeace/Beltrá
78 Reuters/Lucy Nicholson
78 Reuters/Ming Ming
78 TopFoto/ImageWorks

78 Reuters/Rafiquar Rahman
80 Reuters/Viktor Korotayev
81 Reuters/POOL New
82-83 Reuters/Michael Dalder
84-85 Reuters/Jason Reed
86-87 Reuters/Beawiharta Beawiharta
89 Reuters/Christine Muschi

Chapter 4
90-91 Reuters/Chris Helgren
92 Reuters/Rebecca Cook
92 Time Life Pictures/Getty Images
94 Reuters/Str Old
95 Reuters/Peter Morgan
95 Reuters/POOL New
96 Greenpeace
96 Reuters/David Ademas
96 Reuters/Simon Thong
97 Reuters/Miguel Vidal
97 Reuters/Miguel Vidal
97 Reuters/Desmond Boylan
98-99 Reuters/Str Old
101 Reuters/Str Old
101 Greenpeace/Rex C. Curry
102 Reuters/JP Moczulski
102 Reuters/Rebecca Cook
102 Reuters/Rebecca Cook
103 Reuters/Str Old
103 Reuters/George Esiri
104-105 Reuters/Faleh Kheiber

Chapter 5
106-107 Greenpeace/Behring-Chisholm
108 Reuters/Austin Ekeinde
109 Reuters/George Esiri
111 Reuters/Romeo Ranoco
111 Reuters/Supri Supri
111 Reuters/Marcos Brindicci
112 Reuters/Guang Niu
114 Reuters/Andrew Wong
114 Andrew Wong
115 Reuters/China Daily Information Corp-CDIC
116 Greenpeace/Tickle
117 Greenpeace/Martensen
118 Reuters/Andy Soloman
119 Reuters/Rafiquar Rahman
120 Reuters/Amit Gupta
120-121 Jacques Descloitres, MODIS Land Rapid Response Team, NASA/GSFC
122-123 Reuters/Jason Reed

Chapter 6
124-125 Provided by the SeaWiFS Project, NASA/Goddard Space Flight Center, and ORBIMAGE
126 Image courtesy Landsat 7 Science Team and NASA/GSFC
127 Reuters/Steffen Schmidt
129 Andy Harvey
130 National Geographic/Getty Images
131 Greenpeace/Beltrá
131 Topham/Photri
132 Kevin Schafer/Getty Images
133 Grant Gilchrist/Canadian Wildlife Service
134-135 Reuters/Radu Singheti
136-137 Alastair Dobson
138-139 Provided by the SeaWiFS Project, NASA/Goddard Space Flight Center, and ORBIMAGE
140 Reuters/Pawl Kopczynski
141 Reuters/Chor Sokunthea
141 Reuters/China Daily
143 AFP/Getty Images
144-145 Greenpeace/Aslund/Norwegian Polar Institute

Chapter 7
146–147 Reuters/Carlos Barria
148 Greenpeace/Sutton-Hibbert
149 Greenpeace/Sutton-Hibbert
151 Jacques Descloitres, MODIS Land Science Team
152-153 Reuters/Shannon Stapleton
154-155 Reuters/Stefano Rellandini
156 Reuters/Ho Old
156 Reuters/Andy Kuno
156 Reuters/Ho Old
156 Rob Webster
157 Reuters/Alexander Demianchuk
157 Reuters/Adeel Halim

157 Reuters/Kin Cheung
157 Reuters/Nicky Loh
158-159 Getty Images/Ary Diesendruck
160-161 Getty Images/Sakis Papadopoulos
163 His Excellency Mr. Maumoon Abdul Gayoom, President of the Republic of Maldives
163 George Fischer
163 Brian Knutsen
164-165 Jacques Descloitres, MODIS Rapid Response Team, NASA/GSFC
167 Greenpeace/Beltrá

Chapter 8
168-169 TopFoto/ImageWorks
170 Provided by the SeaWiFS Project, NASA/Goddard Space Flight Center
171 Provided by the SeaWiFS Project, NASA/Goddard Space Flight Center
172 Coral Cay Conservation
173 Coral Cay Conservation
174-175 Greenpeace/Grace
176 TopFoto/ImageWorks
177 Topham/Photri
178-179 Jacques Descloitres, MODIS Land Rapid Response Team, NASA/GSFC
180-181 Topfoto/HIP

Chapter 9
182-183 Jeff Schmaltz, MODIS Rapid Response Team, NASA/GSFC
184 Getty Images
185 Reuters/Richard Carson
187 TopFoto/ImageWorks
187 Reuters/Rick Wilking
187 TopFoto/ImageWorks
187 Reuters/Ho New
188 Jacques Descloitres, MODIS Rapid Response Team, NASA/GSFC
188 Jeff Schmaltz, MODIS Rapid Response Team, NASA/GSFC
189 Jeff Schmaltz, MODIS Rapid Response Team, NASA/GSFC
190 Reuters/Lee Celano
190 Reuters/Lee Celano
191 Reuters/Richard Carson
192-193 Jeff Schmaltz, MODIS Rapid Response Team, NASA/GSFC
194 Reuters/Radu Sigheti
194 Kenya Tourist Board
196-197 Greenpeace/Beltrá
198 Greenpeace/Beltrá
199 Greenpeace/Beltrá

Chapter 10
200-201 Reuters/Christine Muschi
202 Reuters/Jose Manuel Ribeiro
203 Reuters/Nacho Doce
204 Reuters/Kimimasa Mayama
204 Reuters/Christine Muschi
206 TopFoto/ImageWorks
207 Topham Picturepoint
208 Reuters/Jason Reed
209 Reuters/Yves Herman
210 Greenpeace/Greig
210 Greenpeace/Weckenmann
211 TopFoto/ImageWorks
211 Langrock/Greenpeace
212-213 Reuters/Stringer Iran
214 Reuters/Keith Bedford
215 Reuters/Pierre Ducharme
216 NOAA Central Library. OAR/ERL/National Severe Storms Laboratory (NSSL)
217 Joseph Golden, NOAA
218 WWF-Canon/Soh Koon Chng
218 TopFoto/ImageWorks
219 CP/Ryan Remiorz
221 Reuters/Christine Muschi
222 Reuters/Tobias Schwarz
222 Reuters/Fred Prouser
223 Greenpeace
225 Reuters/Arko Datta

Chapter 11
226-227 Reuters/China Photo
228 Topham Picturepoint
229 TopFoto
231 Topham Picturepoint
231 Reuters/Claro Cortes
231 Reuters/Reinhard Krause

231 Reuters/Kin Cheung
232 Reuters/Guang Niu
232 WWF-Canon/Soh Koon Chng
233 Reuters/China Daily Information Corp-CDIC
234 Reuters/Str New
234 Reuters/China Daily Information Corp-CDIC
235 Reuters/Guang Niu
236 TopFoto/ImageWorks
238-239 Reuters/Jason Lee
240 TopFoto/ImageWorks
240 Reuters/Issei Kato
241 Reuters/Claro Cortes
242-243 Reuters/Claro Cortes
244 Reuters/China Photo

Chapter 12
246-247 TopFoto/ImageWorks
248 WWF-Canon/Soh Koon Chng
249 Reuters/Alex Grimm
251 TopFoto/ImageWorks
252 Reuters/Ho New
252 Reuters/Stefano Paltera/Solar Decathlon/Handout
255 Reuters/Nir Elias
256-257 Scottish Power
258 Paul West
259 Transport for London
261 European Commission/Influence/M&C Saatchi
261 Chevron
262 BP plc
264 Reuters/Yuriko Nakao
265 Reuters/POOL New
265 Reuters/Toshiyuki Aizawa
266-267 Reuters/Rafiquar Rahman
268 Reuters/Zainal Abd Halim
269 Reuters/Marcos Brindicci
269 Reuters/Paulo Whitaker
270 Greenpeace/Noble
271 Greenpeace/Cannalonga

Chapter 13
272-273 Reuters/Alexei Dityakin
274 TopFoto/UPP
275 Reuters/Robert Pratta
277 Reuters/Str Old
278 Reuters/Mykola Lazarenko/Pool
279 Reuters/Claudia Daut
279 Reuters/Damir Sagolj
279 Reuters/Claudia Daut
280 Paul West
281 Reuters/Christian Charisius
282-283 Reuters/Christian Charisius

Chapter 14
284-285 SES Stirling Energy Systems
286 NASA Goddard Space Flight Center Image by Reto Stöckli
287 NASA Goddard Space Flight Center Image by Reto Stöckli
289 Reuters/Wilson Chu
289 Reuters/Mario Anzuoni
289 Urban Mover
289 General Motors Corporation/Steve Fecht
290 Reuters/Arnd Wiegmann
291 Reuters/Arnd Wiegmann
291 TopFoto/Keystone
292-293 Oddgeir Karlsson
294 Reuters/Reinhard Krause
295 Reuters/China Photo
296-297 Reuters/China Daily Information Corp-CDIC
297 Greenpeace/Jim Hodson
298 Kisho Kurokawa Architects
298 Kiss & Cathcart Architects/Adam Friedberg
300 Nigel Young/Foster and Partners
301 FxFowle Architects/David Sundberg/Esto
302-303 Lifschutz Davidson Sandilands
304 Honda
304 Shell Photographic Services/Shell International Ltd.
305 Reuters/Jess Christensen
306 Richard Lewisohn
306 Greenpeace/Jim Hodson
308 Seafish
308 FARMA/Gareth Jones
311 Reuters/Alberto Lowe
311 Reuters/Kena Betancur

Since I started researching and writing
this book 12 months ago, the amount
of carbon dioxide in the atmosphere
has gone up by four parts per million.
Time is running out.

—Paul Brown
October 2007

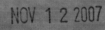